Ernst Schering Research Foundation Workshop 54
The Promises and Challenges
of Regenerative Medicine

Ernst Schering Research Foundation
Workshop 54

The Promises and Challenges of Regenerative Medicine

J. Morser, S. I. Nishikawa
Editors

With 27 Figures

 Springer

Series Editors: G. Stock and M. Lessl

Library of Congress Control Number: 2005925476

ISSN 0947-6075

ISBN-10 3-540-23481-0 Springer Berlin Heidelberg New York
ISBN-13 978-3-540-23481-4 Springer Berlin Heidelberg New York

Springer is a part of Springer Science+Business Media
springeronline.com

© Springer-Verlag Berlin Heidelberg 2005
Printed in Germany

Editor: Dr. Ute Heilmann, Heidelberg
Desk Editor: Wilma McHugh, Heidelberg
Production Editor: Michael Hübert, Leipzig
Cover design: design & production, Heidelberg
Typesetting and production: LE-TEX Jelonek, Schmidt & Vöckler GbR, Leipzig
21/3152/YL – 5 4 3 2 1 0   Printed on acid-free paper

Preface

It has long been known that amphibia and other lower order vertebrates have the capacity to regenerate limbs as well as damaged hearts or brains. Over the past decade, there has been a major change in the way that the potential for regeneration in mammals is viewed. Earlier, in contrast to the acceptance of regeneration in amphibia, it was generally believed that there was very limited if any capacity for regeneration in many mammalian organ systems such as the heart and brain. The discovery of tissue-resident adult stem cells and the description of the properties of embryonic stem cells have altered this view. This change in paradigm

has led to the hope that these discoveries can be harnessed in medical practice to cure chronic disabling diseases.

The use of tissue-resident adult stem cells depends on the ability to either mobilize them or to convert them from one lineage to another. These problems do not arise with embryonic stem cells. Instead, their use is fraught with ethical and political issues as well as the question of how to direct their differentiation toward the desired cell type. Whichever approach is taken, issues of safety have to be paramount. In particular, the role of stem cells in tumorigenesis is critical in assessing their potential clinical utility.

The Ernst Schering Research Foundation and the Riken Center on Developmental Biology jointly organized a workshop on "The Promises and Challenges of Regenerative Medicine," which took place in Kobe, Japan on 20–22 October 2004. The purpose of the workshop was to discuss the present state of knowledge and future directions in this important field. Leading basic scientists and clinicians reviewed and discussed several timely topics within three main themes: (1) evolution, development, and regeneration, including stem cells in Planaria and stem cell niches; (2) embryonic and adult stem cells, including a discussion of the regulatory system in Japan for human embryonic stem cells; and (3) regeneration in specific indications including a discussion of the role of stem cells in organs such as the skin, brain, liver, pancreas, cornea, and the cardiovascular system. In addition, the role of stem cells in glioblastoma was presented along with the implications for other tumors.

This book contains the proceedings of the workshop. The individual contributions give an excellent overview of the basic cellular biology and clinical aspects of this emerging field. In addition to summarizing the present state of the art of regenerative medicine, this book also points out some of the basic unanswered questions where future research is needed before these approaches can be introduced into routine clinical practice.

Shin Ichi Nishikawa,
John Morser

Contents

List of Editors and Contributors

Editors

Morser, J.
Nihon Schering K.K. Research Center, BMA 3F,
1-5-5- Minatojima-Minamimachi, Chuo-ku, Kobe 650-0047, Japan
(e-mail: john_morser@berlex.com)

Nishikawa, S.I.
The Center For Developmental Biology,
2-2-3 Minatojima-Minamimachi, Chuo-ku, Kobe 650-0047, Japan
(e-mail: nishikawa@cdb.riken.jp)

Contributors

Asahara, T.
Riken Center for Developmental Biology,
2-2-3 Minatojima-Minamimachi, Chuo-Ku, 650-0047 Kobe, Japan
(e-mail: asa777@aol.com)

Ghadially, R.
Department of Dermatology, University of California, San Francisco,
VA Medical Center (190), 4150 Clement St., San Francisco, CA 94121 USA
(e-mail: rghadial@itsa.ucsf.edu)

Iwama, A.
Laboratoy of Stem Cell Therapy, Center for Experimental Medicine,
The Institute of Medical Science, University of Tokyo,
4-6-1 Shirokanedai, Minato-ku, Tokyo, Japan 108-8639

Kinoshita, S.
Department of Ophthalmology, Kyoto Prefectural University of Medicine,
Kawaramachi-hirokoji, Kamigyo-ku, Kyoto 602-0841, Japan
(e-mail: shigeruk@ophth.kpu-m.ac.jp)

Nakamura, T.
Department of Ophthalmology, Kyoto Prefectural University of Medicine,
Kawaramachi-hirokoji, Kamigyo-ku, Kyoto 602-0841, Japan

Nakatsuji, N.
Department of Development and Differentiation,
Institute for Frontier Medical Sciences, Kyoto University,
53 Kawahara-cho Shogoin, Sakyo-ku, 606-8507 Kyoto, Japan
(e-mail: nnakatsu@frontier.kyoto-u.ac.jp)

Nakauchi, H.
Laboratory of Stem Cell Therapy, Center for Experimental Medicine,
Institute of Medical Science, University of Tokyo,
4-6-1, Shirokane-dai, 108–8639 Minato-ku, Tokyo, Japan

Negishi, M.
Laboratory of Stem Cell Therapy, Center for Experimental Medicine,
The Institute of Medical Science, University of Tokyo,
4-6-1 Shirokanedai, Minato-ku, Tokyo, Japan 108-8639

Nishikawa, S.I.
Riken Center for Developmental Biology,
2-2-3 Minatojima-Minamimachi, Chuo-Ku, 650-0047 Kobe, Japan
(e-mail: nishikawa@cdb.riken.jp)

Oguro, H.
Laboratory of Stem Cell Therapy, Center for Experimental Medicine,
The Institute of Medical Science, University of Tokyo,
4-6-1 Shirokanedai, Minato-ku, Tokyo, Japan 108-8639

Osawa, M.
Riken Center for Developmental Biology,
2-2-3 Minatojima-Minamimachi, Chuo-Ku, 650-0047 Kobe, Japan

Sasai, Y.
Riken Center for Developmental Biology,
2-2-3 Minatojima-Minamimachi, Chuo-Ku 650-0047 Kobe, Japan
(e-mail: sasai@cdb.riken.jp)

Trounson, A.
Monash Immunology and Stem Cell Laboratories,
Monash University and Australian Stem Cell Centre, Wellington Road,
Clayton, Victoria, 3800, Australia
(e-mail: jillian.mcfadyean@med.monash.edu.au)

1 Melanocyte System for Studying Stem Cell Niche

S.I. Nishikawa, M. Osawa

Abstract. There are many notions in stem cell biology that lack proof. The stem cell niche is the most typical example. While it is a convenient terminology for designating anything that supports stem cells, the cellular basis of the niche is poorly understood for many stem cell systems. In this chapter, we describe how useful the melanocyte system would be for investigating the molecular and cellular basis of the niche.

It has been difficult to find a common definition of stem cell systems, as it is found from the most primitive multicellular organisms such as Volvox up to the highly complex organisms such as mammals. Stem cell (SC) systems in mammals are often characterized as a system containing three components, immature stem cells compartment(SCC that are quiescent in proliferation, a rapidly proliferating compartment called the transit amplifying stage (TA), and the differentiating compartment (DC). However, there are also many stem cell systems in which the quiescent SC is difficult to find and most immature cells are actively cycling. A good example of this is planaria, in which all somatic cells are recruited from constitutively renewing stem cells. In planaria, only one type of somatic stem cells appear to exist, but they are multipotent, recruiting all cell lineages including germ cells (Agata 2003). On the other hand, for researchers of Drosophila, stem cell systems are unipotent stem cell systems in most cases (Spradling et al. 2001). Despite such a striking diversity among stem cell systems, the word "niche" is used preferably by most stem cell researchers to describe the mechanism regulating stem cell activity.

We believe that the reason for this term gaining such popularity relates to the role of the stem cell system in renewing tissue cells according to the body's requirements. In the stem cell system, cell recruitment should be controlled strictly according to the demand of the tissues and body. Either low production or overproduction of progenies is fatal for the organism. Thus it is conceivable that such a demand is represented somewhere in the environmental architecture in the close vicinity of stem cells. Moreover, the term "niche" is abstract enough to be exploited for diverse situations, but also concrete enough to express an image that can be readily understood. Thus, niche can represent all possible beings as an environment for stem cells.

Despite popular awareness of the importance of niche in the regulation of the stem cell system, niche is still more concept than reality in many stem cell systems. I would emphasize that molecular and cellular characterization of the niche has just started using the gonadal tissues of Drosophila and the nematode (Spradling et al. 2001), whereas it is still an open question for stem cell systems of higher organisms.

1.1   Advantages of Melanocyte Stem Cell System for Investigating the Stem Cell Niche

Among possible architectures for the stem cell system, the most common one in vertebrates is the one consisting of three compartments, SC, TA, and DC. In this situation, niche is defined as the environment that regulates and switches between quiescent SC and activated TA states in the stem cell system. Thus, stem cell research in mammals normally starts from distinguishing the SC from other compartments. If the SC is specified successfully, then the niche can be defined as the environment the surrounding SC. In fact, the distinction between the SC and the other compartments is still the major subject for many stem cell systems, though there are also those that have been characterized to a great extent. Among such well-characterized stem cell systems, we believe that the melanocyte stem cell system provides an ideal model for investigating stem cell niche for following reasons.

In the mouse, most melanocytes reside in hair follicles. Thus, their environment comprises only the components present within the hair follicle. More importantly, in the melanocyte system, the SC and other compartments are segregated from each other in the different regions of hair follicles. Namely, the SC resides just below the bulge region (sub-bulge region: SBR), whereas other compartments are present in the matrix region of hair follicles (Fig. 1). Hence, compared with most stem cell systems where the SC and the TA exist in the same region, it is easy to

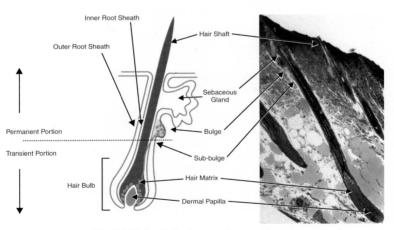

Fig. 1. Structure of hair follicle. Only the transient portion undergoes the regeneration cycle. Stem cells for keratinocytes are shown to be present in the bulge area, whereas those for melanocyte are in the sub-bulge area

distinguish the SC from other compartments. This spatial segregation of the SC and other compartments is also important to framing the structure of the niche. In most stem cell systems, the SC and its immediate progenies the TA exist side by side (Fig. 2). In this situation, the TA can be regarded as a component of the environment for the SC as well as SC's progenies. Indeed, there is an extreme hypothesis that the SC by itself produces its own niche in such a way that the SC promotes neighboring cells to commit to the TA, thereby inhibiting their reversion to the SC, whereas the TA generated in the neighborhood of the SC plays a role of the niche. This idea indeed challenges the naïve thinking that separates the stem cell components from environmental components. Of note is that accumulating evidence suggests that keratinocytes in the epidermis as well as in the hair follicle provide such an example (Watt 1998; Blapain et al. 2004). Nonetheless, in most stem cell systems where

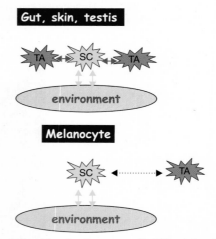

Fig. 2. Cellular composition in the stem cell (*SC*) niche. In most stem cell systems, SCs are located together with both the TA of the same lineage and surrounding cells of the other lineages. In this situation, it is difficult to rule out the involvement of the TA in the niche function. On the other hand, in the melanocyte stem cell system, SCs and the TA are segregated completely. Thus, the niche is formed only by the cells of lineages distinct from melanocyte

eventually become pigmented in two to three regeneration cycles. This result suggests not only the plasticity of differentiated melanocytes, but also the dominant role of the niche in inducing the reversion to the SC.

A dominant role of the sub-bulge niche in directing SC fate was confirmed by analyzing the neonatal process of melanocyte colonization to newly formed hair follicles. By a series of immunohistochemical studies, we noted that Sox10 that is requisite for melanocyte development is downregulated in the SC in the sub-bulge region, whereas it is expressed in all melanocytes throughout embryonic life and also those in the hair matrix (data not shown). Taking advantage of this difference in the Sox10 expression between the SC and other compartments, we investigated the Sox10 expression during the neonatal process from the stage when melanocytes enter the developing follicle to the stage when some of them eventually settled in the sub-bulge niche. Indeed, nearly all melanocytes are Sox10$^+$ at the time when they first enter developing hair follicles. However, when they pass through the sub-bulge region, a small portion of melanocytes downregulate Sox10, whereas most of those that pass this region further down to the hair matrix region maintain Sox10 expression. This observation strongly suggests that only a special condition present in the SB region can direct melanocytes to become SCs. Our observation that only a small portion of melanocytes receive the niche signal to downregulate Sox10, whereas many more melanocytes pass through the sub-bulge region and reach the hair matrix suggests that the actual niche is present in a further limited area of the SB region. Moreover, if the range where the effect of the niche is reached is short, the niche function may be blocked by the colonization of the SC to a niche. Alternatively, it is possible that the niche activity is established only after some melanocytes have passed this region. Nonetheless, the two experiments indicate that the niche plays a dominant role in directing activated melanocytes to become SCs.

1.4 Melanocyte Life Cycle

What resulted from our phenomenological analysis of melanocyte SCs in the hair follicles is that the niche is the environment that plays two roles, in the induction of SC state on one hand and in the maintenance of the survival of quiescent melanocytes in the absence of SCF on the other.

Based on these new understandings of the role of the niche for melanocyte SC, we summarized the life cycle of melanocytes in the hair follicle in Fig. 6. Melanocytes in the epidermis of neonates are in the activated state. While SCF is not expressed in the epidermis of adult mice, it is expressed in that of neonatal mice. This neonatal expression of SCF is essential for supporting the process of melanocyte colonization into hair follicles, blocking c-Kit function at neonatal stage depletes most melanocytes in the skin (Nishimura et al. 2002). Of note is that the maturation of hair follicle occurs in a successive wave for 3–4 days after birth. In this maturation process, hair follicles elongate and the niche is newly formed. With the c-Kit block experiment, it was confirmed that niche formation completed only 20% of hair follicles and indeed it takes 3–4 days for all follicles to establish SC–niche interaction (Nishikawa et al. 1991). Along with this maturation process, the SCF expression in the skin is downregulated gradually, except in the hair papilla. This downregulation results in the death of melanocytes in the interfollicular epidermis and also in most hair follicles. Nonetheless, niches are formed

Life cycle of melanocyte

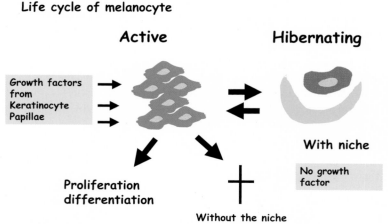

Fig. 6. Life cycle of the melanocyte stem cell system. For details, see text

before all melanocytes die out from the sub-bulge area, eventually allowing a few melanocytes to establish interaction with the niche. After this neonatal process, melanocytes are found only in the SB and the hair matrix and repeat the regeneration cycle. During the regeneration cycle, the melanocytes in the matrix region undergo apoptosis along with that of follicular keratinocytes and the new melanocytes are replenished from the SC remaining in the sub-bulge region. Proliferation of melanocytes is allowed only in the matrix region because it is the sole region where SCF is available together with other growth factors, whereas they die out in other SCF-free regions. On the other hand, the SB niche can provide an environment that allows their long-term survival in the absence of SCF. We believe that the transition from active to quiescent states is also induced by the niche, though it is not clear whether the two processes are regulated by the same cells. This view of the melanocyte life cycle in terms of its relation to the niche is illustrated in Fig. 5. We want to emphasize that this life cycle of the melanocyte is reminiscent to the hibernation/awakening cycle of animals in many respects. In both cases, hibernation is necessary because the factors required for the maintenance of the active state are absent. Indeed, death is inevitable if there is no hibernation. Then the preparatory state for hibernation such as depression of the general metabolic activity is induced actively, and finally the actual hibernation starts. Therefore, we designated this cycle as the hibernation cycle of the stem cell and illustrated in Fig. 6.

1.5   Toward Elucidation of Molecular Mechanisms Underlying Niche–SC Interaction

Everything we have described so far is only phenomenology, and nothing is known about the molecules involved in establishing and maintaining the SC–niche relationship. For this purpose, isolation of both the SC and the niche is essential. We have established two methods to isolate the SC specifically. One is the method using FACS and the other is single cell isolation. Both methods are based on a transgenic mouse strain in which melanocytes express a high level of GFP. Using the same materials and methods, melanocytes at other stages are purified both from embryos and adult mice. After purifying SC and other compartments of the melanocyte stem cell system, we compared the gene expression profile of two populations using various methods, including DNA array.

In contrast to melanocytes themselves, it is difficult to isolate a specific niche for melanocyte SC, because of lack of molecular markers for the niche. The cellular composition of the sub-bulge region suggests that, if niche cells exist, they should be keratinocytes. However, how the niche cells are different from those in other regions of the hair follicle is totally unknown. Only one hint that we have found so far is the presence of keratinocytes that form clusters with the melanocyte SC during dissociation of hair follicles. Thus, although it is not clear if those cells correspond to the niche cells, we are isolating them separately as the niche cells. Given that there are no consistent markers for defining the niche cells, there may not be a choice other than pursuing this possibility. Hence, even though we are able to purify the SC and other compartments of melanocytes by FACS, the method to isolate them manually under the microscope may be valuable.

References

Agata K (2003) Regeneration and gene regulation in planarians. Curr Opin Genet Dev 13:492–496

Blanpain C, Lowry WE, Geoghegan A, Polak L, Fuchs E.(2004) Self-renewal, multipotency, and the existence of two cell populations within an epithelial stem cell niche. Cell 118:635–648

Hibberts NA, Messenger AG, Randall VA (1996) Dermal papilla cells derived from beard hair follicles secrete more stem cell factor (SCF) in culture than scalp cells or dermal fibroblasts. Biochem Biophys Res Commun 222:401–405

Jahoda CA, Horne KA, Oliver RF (1984) Induction of hair growth by implantation of cultured dermal papilla cells. Nature 311:560–562

Kunisada T, Yoshida H, Yamazaki H, Miyamoto A, Hemmi H, Nishimura E, Shultz LD, Nishikawa S, Hayashi S (1998) Transgene expression of steel factor in the basal layer of epidermis promotes survival, proliferation, differentiation and migration of melanocyte precursors. Development 125:2915–2923

Nishikawa S, Kusakabe M, Yoshinaga K, Ogawa M, Hayashi S, Kunisada T, Era T, Sakakura T (1991) In utero manipulation of coat color formation by a monoclonal anti-c-kit antibody: two distinct waves of c-kit-dependency during melanocyte development. EMBO J 10:2111–2118

Nishimura EK, Jordan SA, Oshima H, Yoshida H, Osawa M, Moriyama M, Jackson IJ, Barrandon Y, Miyachi Y, Nishikawa S (2002) Dominant role of the niche in melanocyte stem-cell fate determination. Nature 416:854–860

Spradling A, Drummond-Barbosa D, Kai T (2001) Stem cells find their niche. Nature 414:98–104

Watt FM (1998) Epidermal stem cells: markers, patterning and the control of stem cell fate. Philos Trans R Soc Lond B Biol Sci 353:831–837

2 Establishment and Manipulation of Monkey and Human Embryonic Stem Cell Lines for Biomedical Research

N. Nakatsuji

Abstract. We have established several embryonic stem (ES) cell lines of the cynomolgus monkey. They maintain the normal karyotype and pluripotency in culture for long periods. We obtained government approval and grants to produce human ES cell lines from frozen surplus embryos in April 2002. We have established and characterized three human ES cell lines (KhES-1, KhES-2, KhES-3). We started distribution of human ES cells to other research groups in March 2004. It would be important to produce genetically modified monkey and human ES cells for various purposes. After improvement of the transfection and selection methods, we have produced monkey ES cells with integrated transgenes at efficient and reliable rates. We are also investigating reprogramming of somatic cells into pluripotent stem cells by cell fusion with ES cells. Such reprogramming could be used to produce pluripotent stem cells for each patient without therapeutic cloning, which would raise ethical concerns.

2.1 Introduction

For many years, mouse embryonic stem (ES) cell lines have been important tools for basic biology to investigate cell differentiation and development in mammals, as well as to carry out gene targeting for analysis of gene function in gene-disrupted knock-out mice. Establishment of human ES cell lines in 1998, from surplus human embryos produced at infertility clinics, has indicated great potential of ES cells in medical research and application such as cell therapy and drug discovery (Thomson et al. 1998; Reubinoff et al. 2000) (Fig. 1).

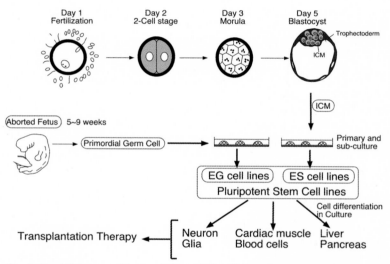

Fig. 1. Establishment of human ES and EG cell lines from blastocysts and its medical application. Development of preimplantation stage human embryos, isolation of the inner cell mass, and establishment of ES cell lines are illustrated. ES cells can be induced to differentiate into various cell types for cell therapy or other application. Derivation of embryonic germ (*EG*) cell lines, another pluripotent cell line, from fetal germ cells is also shown

ES cells have two important abilities: unlimited rapid proliferation enabling production of large numbers of human cells and pluripotency to differentiate into almost any type of cell. Such abilities also enable genetic alteration of ES cells, including insertion of various gene constructs designed for particular purposes and also gene targeting to modify endogenous genes by homologous recombination.

In order to advance medical applications using human ES cells, it is very important to devise reliable methods to establish and maintain ES cells in defined conditions, and also to discover various methods of inducing their differentiation into specific cell types. Already, there have been reports of the production of various types of neurons, glia, cardiac muscle, hematopoietic cells, and endothelial cells, all of which have important medical applications.

2.2   Cynomolgus Monkey ES Cell Lines

2.2.1   Establishment and Characterization of Monkey ES Cell Lines

We have established several ES cell lines from blastocysts of cynomolgus monkey (Suemori et al. 2001) (Fig. 2). ES cell lines have been established from nonhuman primates including rhesus monkey (*Macaca mulatta*), common marmoset (*Callithrix jacchus*) and cynomolgus monkey (*Macaca fascicularis*) (Thomson et al. 1995, 1996; Suemori et al. 2001). These primate ES cell lines have very similar characteristics, including with human ES cell lines. They express alkaline phosphatase activity and stage-specific embryonic antigen (SSEA)-4 and, in most cases, SSEA-3. Their pluripotency is confirmed by the formation of embryoid bodies and differentiation into various cell types in culture and also by the formation of teratomas that contained many types of differentiated tissues, including derivatives of three germ layers after transplantation into immunodeficient SCID mice.

When compared to mouse ES cells, the noneffectiveness of the leukemia inhibitory factor (LIF) in maintenance of stem cells makes culture of primate and human ES cell lines difficult and prone to undergoing spontaneous differentiation. Also, these ES cells are more susceptible to various stresses, causing difficulty with subculturing using enzymatic

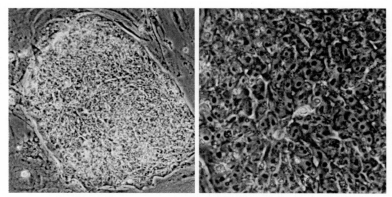

Fig. 2. ES cell lines established from blastocysts of cynomolgus monkey. Phase contrast micrographs of the monkey ES cell colonies at lower (*left*) and higher (*right*) magnification are shown

treatment, cloning from single cells, and gene transfection. However, with various improvements in culture methods (Suemori and Nakatsuji 2003), it is possible to maintain stable colonies of monkey ES cells using a serum-free medium and subculturing with trypsin treatment. Under such conditions, we can maintain cynomolgus monkey ES cell lines in an undifferentiated state with a normal karyotype and pluripotency even after prolonged periods of culture over 1 year (Suemori et al. 2001;

Table 1. Karyotype analysis of two cynomolgus monkey ES cell lines

Cell line	Passage (month)	Normal/counted (%)	Sex
CMK6	47 (6)	14/20 (70%)	Male
	84 (12)	17/25 (68%)	
CMK9	18 (3)	16/20 (80%)	Female
	64 (7)	14/16 (88%)	

One male (CMK6) and one female (CMK9) cell line were grown and passaged for several months and their karyotypes were examined. All the chromosome preparations at mitosis were counted and the majority showed the normal karyotype

Suemori and Nakatsuji 2003) (Table 1). Such monkey ES cells can be induced to differentiate various cell types in culture (Kawasaki et al. 2002; Mizuseki et al. 2003; Umeda et al. 2004; Haruta et al. 2004).

2.2.2 Significance of Monkey ES Cells

Nonhuman primate ES cell lines provide important research tools for basic and applied research. First, they provide wider aspects of investigation of the regulative mechanisms of stem cells and cell differentiation among primate species. Second, their usage can be valuable preparation for research using human ES cells, which is under strict ethical regulation in many countries. Last and most important, they are indispensable for animal models of cell therapy to test effectiveness, safety, and immunological reaction of the allogenic transplantation in a setting similar to the treatment of human diseases.

The safety of cell transplantation must be tested using animal models before clinical application of ES cells. ES cells are not malignant tumor cells, but they show unlimited proliferation and form benign tumors in immunorepressed animals. Therefore, removal of stem cells from differentiated cell populations is necessary before cell transplantation. For even a higher degree of safety, implementation of a suicidal or cell ablation gene in ES cells may be necessary to trigger cell death if something goes wrong after transplantation.

Another important aspect is the immunological response after transplantation of allogenic ES-derived cells, or possibly, after genetic alteration of ES cell lines to reduce their antigenicity. The major histocompatibility (MHC) genes of ES cells could be altered by gene targeting and subsequent introduction of desired MHC-type genes. Complete matching of the many MHC types would be impossible between the modified ES cells and recipients, but matching of the major types might reduce immunological rejection to clinically manageable levels. For such immunological tests, nonhuman primates provide the best allogenic combination of ES cell lines and disease model animals.

For preclinical research of actual cell transplantation, it is also important that the sizes and structures of various organs and tissues in animal models are similar to those of humans. This is especially important for surgical procedures; the absolute size and depth of structures in the brain, for example, determine the cell number and transplantation protocol that needs to be evaluated.

Among the monkeys, macaques such as the rhesus and cynomolgus monkeys are most suitable as model nonhuman primates. They are bred as experimental animals and widely used for medical research. Also, various disease models in macaques are available for research purposes. For such reasons, we established cynomolgus monkey ES cell lines and devised improved methods for culture and manipulation of monkey ES cells.

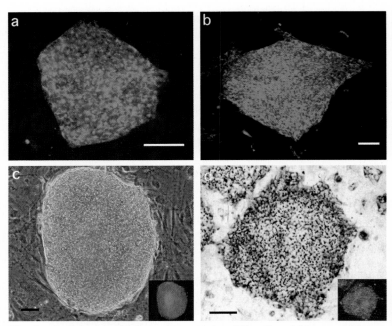

Fig. 3a–d. Colonies of cynomolgus monkey ES cells, which are stably expressing the EYFP-mito fluorescent protein gene. After gene transfection with electroporation and sorting with G418-registance, all the ES cells in the colony express the fluorescent protein uniformly. They also express stem cell markers, alkaline phosphatase, and SSEA-4 antigen

2.2.3   Genetic Alteration of Monkey ES Cells

For further progress in basic research and the medical application of primate and human ES cells, we need to improve many aspects of the manipulation of stem cells. For example, we would like to produce genetically modified monkey and human ES cells for reduction of anti-genicity in cell therapy or efficient selection of particularly useful cell types. After improvement of the transfection and selection methods, we can produce ES cell clones with integrated transgenes at efficient and reliable rates (Furuya et al. 2003) (Figs. 3, 4). However, we have experienced significant differences among monkey ES cell lines in efficiency of transfection and sensitivity to the electroporation procedure. Such manipulation methods for modification of primate ES cells will be utilized to produce new cell lines improved for particular purposes or analyses of molecular and cellular mechanisms related to pluripotency and differentiation.

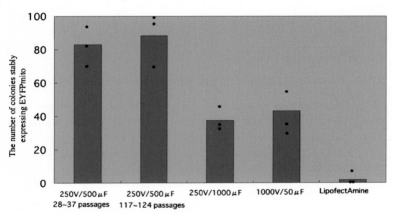

Fig. 4. Conditions of electroporation and transfection efficiency to produce colonies with stable integration of the introduced gene. 1×10^7 cynomolgus monkey ES cells (CMK6) were used for each experiment. Adequate conditions yielded 80 or more G418-registant colonies that expressed the transfected gene

2.2.4   STAT3 Signaling and Pluripotency in Monkey ES Cells

The signal transducer and activator of transcription 3 (STAT3) pathway,
which can be activated by leukemia inhibitory factor (LIF), plays an
essential role in the maintenance of self-renewal and pluripotency in
mouse ES cells. In human and monkey ES cells, however, LIF is not
sufficient to maintain pluripotency and the undifferentiated state. We
have investigated significance of the STAT3 signaling pathway in monkey ES cells by making use of improved transfection methods (Furuya
et al. 2003). We found that stimulation of cynomolgus monkey ES cells
with LIF or IL-6/sIL-6R leads to STAT3 phosphorylation, as in murine
ES cells. In addition, nuclear translocalization and transcriptional activation of STAT3 occurred in a LIF-dependent manner. Moreover, the
analysis of a dominant interfering mutant, STAT3F, showed that monkey
ES cells continued to proliferate in the undifferentiated state, retaining
their pluripotency when the phosphorylation, nuclear translocalization,
and transcriptional activation of endogenous STAT3 following LIF stimulation were completely abrogated by overexpressing STAT3F. These
results demonstrated that the STAT3 pathway functions in cynomolgus
monkey ES cells but is not essential for the maintenance of pluripotency
(Sumi et al. 2004).

2.3   Human ES Cell Lines

2.3.1   Guidelines for the Human ES Cell Research in Japan

We are now carrying out a project to establish human ES cell lines from
frozen surplus human embryos. The Japanese government guidelines
for the human ES cell research, issued in September 2001, require strict
procedures for establishment and research usage of human ES cells, as
shown in Table 2.

2.3.2   Establishment and Characterization of Human ES Cell Lines

We obtained the approval and national grants to produce human ES
cell lines in April 2002. So far, we have established and characterized
three human ES cell lines (KhES-1, KhES-2, KhES-3) (Fig. 5). We are

Table 2. Japanese guidelines on derivation and research usage of human ES cell lines

1. Human embryos that can be used for derivation of ES cells:
   – Frozen embryos produced for fertility treatments
   – After decision made by genetic parents not to be used for uterine transfer
   – Informed consent from donor couples with 1 month waiting period before use
2. Research institute to establish ES cell lines:
   – Enough experience in establishment of animal ES cell lines
   – Obligation to distribute ES cells to other institutes with approved research plans
3. Research institute to use ES cell lines:
   – Research for advancement of life science, medicine, and human health
   – Scientific necessity to use human ES cells in research plans
   – Guidelines for clinical usage are now being drawn up by the Ministry of Health
4. Reviewing and approval of research plans:
   – Proposal of derivation or usage of human ES cells must be approved by the institutional review board and also by the government committee

Only the principles are shown. The research institute to establishment human ES cell lines can use only the frozen embryos left over from fertility treatment after proper steps are taken to obtain informed consent for donation of the embryos. It also has the obligation to distribute the established ES cell lines to the research institute that obtained government approval of their research plans to make use of the ES cells for medical and related research

required to distribute ES cells to other institutes in Japan with approved research plans, and we have started the distribution in March 2004.

2.4 Reprogramming of Somatic Cells by Fusion With ES Cells

Immunological rejection is the major problem for clinical application of cell therapy using ES cells. There are several possible ways to avoid the rejection (Table 3). They include genetic alteration in MHC genes and therapeutic cloning. We are investigating reprogramming of somatic cells into pluripotent stem cells by cell fusion with ES cells (Tada et al.

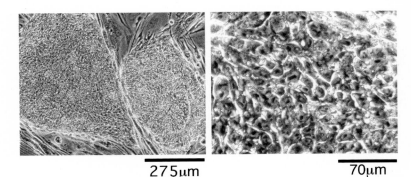

275μm   70μm

Fig. 5. One of three human ES cell lines established in Japan (KhES-1). A lower (*left*) and higher (*right*) magnification phase contrast micrographs are shown. So far, the authors have established three cell lines, KhES-1, KhES-2, and KhES-3. They are growing stably in culture and are ready to be distributed to other institutes in Japan

Table 3. Possible options to obtain stem cells that can avoid immunological rejection in cell therapy

1. Genetic alteration of ES cells by gene targeting of MHC genes to reduce immunogenicity

2. Reprogramming of somatic cells:

   a. By nuclear transfer into oocytes to produce blastocysts and then ES cells

   b. By cell fusion with ES cells

   c. By treatment with reprogramming factors after mechanisms are elucidated in future

Genetic alterations such as disruption of the major histocompatibility genes (HLA genes) in ES cells could reduce antigenicity of the differentiated cells in cell therapy. However, a perfect solution may need reprogramming of somatic cells from patients into pluripotent stem cells by nuclear transfer into oocyte, cell fusion with ES cells, or treatment with reprogramming factor(s) in the future. The last option may be the best but probably requires the longest research period before clinical application

2001, 2003; Kimura et al. 2004). We have found that the nucleus and genomic function of somatic cells are transformed into those of the pluripotent cells when fused with ES or EG cells. Such reprogramming could be utilized to produce pluripotent stem cells that have the same genome with a patient to avoid immunological rejection without production of the nuclear-transfer-cloned embryos, which would raise various ethical problems in many countries, including Japan.

Acknowledgements. The human ES cell project has been supported by the National Bio-Resource Project of the Ministry of Education, Culture, Sports, Science, and Technology.

References

Furuya M, Yasuchika K, Mizutani K, Yoshimura Y, Nakatsuji N, Suemori H (2003) Electroporation of cynomolgus monkey embryonic stem cells. Genesis 37:180–187

Haruta M, Sasai Y, Kawasaki H, Amemiya K, Ooto S, Kitada M, Suemori H, Nakatsuji N, Ide C, Honda Y, Takahashi M (2004) In vitro and in vivo characterization of pigment epithelial cells differentiated from primate embryonic stem cells. Invest Ophth Vis Sci 45:1020–1025

Kawasaki H, Suemori H, Mizuseki K, Watanabe K, Urano F, Ichinose H, Haruta M, Takahashi M, Yoshikawa K, Nishikawa SI, Nakatsuji N, Sasai Y (2002) Generation of dopaminergic neurons and pigmented epithelia from primate ES cells by stromal cell-derived inducing activity. Proc Natl Acad Sci U S A 99:1580–1585

Kimura H, Tada M, Nakatsuji N, Tada T (2004) Histone code modifications on pluripotential nuclei of reprogrammed somatic cells. Mol Cell Biol 24:5710–5720

Mizuseki K, Sakamoto T, Watanabe K, Muguruma K, Ikeya M, Nishiyama A, Arakawa A, Suemori H, Nakatsuji N, Kawasaki H, Murakami F, Sasai Y (2003) Generation of neural crest-derived peripheral neurons and floor plate cells from mouse and primate embryonic stem cells. Proc Nat Acad Sci U S A 100:5828–5833

Reubinoff BE, Pera MF, Fong CY, Trounson A., Bongso A (2000) Embryonic stem cell lines from human blastocysts: somatic differentiation in vitro. Nat Biotechnol 18:399–404

Suemori H, Nakatsuji N (2003) Growth and differentiation of cynomolgus monkey ES cells. Methods in enzymology (vol. 365) In Wassarman PM,

Keller GM (eds) Differentiation of embryonic stem cells. Elsiever, San Diego, pp 419–429

Suemori H, Tada T, Torii R, Hosoi Y, Kobayashi K, Imahie H, Kondo Y, Iritani A, Nakatsuji N (2001) Establishment of embryonic stem cell lines from cynomolgus monkey blastocysts produced by IVF or ICSI. Dev Dynamics 222:273–279

Sumi T, Fujimoto Y, Nakatsuji N, Suemori H (2004) STAT3 is dispensable for maintenance of self-renewal in nonhuman primate embryonic stem cells. Stem Cells 22:861–872

Tada M, Takahama Y, Abe K, Nakatsuji N, Tada T (2001) Nuclear reprogramming of somatic cells by in vitro hybridization with ES cells. Curr Biol 11:1553–1558

Tada M, Morizane A, Kimura H, Kawasaki H, Ainscough JFX, Sasai Y, Nakatsuji N, Tada T (2003) Pluripotency of reprogrammed somatic genomes in ES hybrid cells. Dev Dynamics 227:504–510

Thomson JA, Kalishman J, Golos TG, Durning M, Harris CP, Becker RA, Hearn JP (1995) Isolation of a primate embryonic stem cell line. Proc Natl Acad Sci U S A 92:7844–7848

Thomson JA, Kalishmanm J, Golos TG, Durning M, Harris CP, Hearn JP (1996) Pluripotent cell lines derived from common marmoset (Callithrix Jacchus) blastocysts. Biol Reprod 55:254–259

Thomson JA, Itskovitz-Eldor J, Shapiro SS, Waknitz MA, Swiergiel JJ, Marshall VS, Jones JM (1998) Embryonic stem cell lines derived from human blastocysts. Science 282:1145–1147

Umeda K, Heike T, Yoshimoto M, Shiota M, Suemori H, Luo HY, Chui DHK, Torii R, Shibuya M, Nakatsuji N, Nakahata T (2004) Development of primitive and definitive hematopoiesis from nonhuman primate embryonic stem cells in vitro. Development 131:1869–1879

3 Human Embryonic Stem Cell Derivation and Directed Differentiation

A. Trounson

Abstract. Human embryonic stem cells (hESCs) are produced from normal, chromosomally aneuploid and mutant human embryos, which are available from in vitro fertilisation (IVF) for infertility or preimplantation diagnosis. These hESC lines are an important resource for functional genomics, drug screening and eventually cell and gene therapy. The methods for deriving hESCs are well established and repeatable, and are relatively successful, with a ratio of 1:10 to 1:2 hESC lines established to embryos used. hESCs can be formed from morula and blastocyst-stage embryos and from isolated inner cell mass cell (ICM) clusters. The hESCs can be formed and maintained on mouse or human somatic cells in serum-free conditions, and for several passages in cell-free cultures. The hESCs can be transfected with DNA constructs. Their gene expression profiles are being described and immunological characteristics determined. They may be grown indefinitely in culture while maintaining their original karyotype but this must be confirmed from time to time. hESCs spontaneously differentiate in the absence of the appropriate cell feeder layer, when overgrown in culture and when isolated from the ESC colony. All three major embryonic lineages are produced in differentiating attachment cultures and in unattached embryoid bodies. Cell progenitors of interest can be identified by markers, expression of reporter genes and characteristic morphology, and the culture thereafter enriched for further culture to more mature cell types. The most advanced directed differentiation pathways have been developed for neural cells and cardiac muscle cells, but many other cell types including haematopoietic progenitors, endothelial cells, lung alveoli, keratinocytes, pigmented retinal epithelium, neural crest cells and motor neurones, hepatic progenitors and cells that have some markers of gut tissue and pancreatic cells have been produced. The prospects for regenerative medicine are significant and there is much optimism for their contribution to human medicine.

3.1   Introduction

Stem cells are clonogenic, self-renewing progenitor cells that are able to produce one or several more specialised cell types. They are arbitrarily classified as embryonic stem cells (ESCs) or adult stem cells (ASCs), the latter being found in mature tissues of the embryo, fetus or postnatal organism. ASCs are often slowly replicating (RBdU label-retaining cells) that are present in specialised niches that control their mobilisation and differentiation pathways. ESCs, as distinct from ASCs, are immortal and pluripotent, and are capable of forming all the tissues of the body (Pera et al. 2000). Embryonic germ cells (EGCs) may be formed from the primitive gonadal ridges of the developing embryo or fetus (human 6–9 weeks of gestation) and have many of the pluripotential properties of ESCs (Shamblott et al. 1998). Few ASCs have the plasticity for differentiation that ESCs have, most are uni- or multipotential and usually form only the cells of the lineage of the tissue they originate from. However, some ASCs, such as bone marrow-derived stem cells, will exhibit plasticity for colonising a variety of tissues under some experimental situations, and in response to tissue damage and inflammation (Herzog et al. 2003). There are some bone marrow-derived cells that can be selected in the laboratory that demonstrate multiplicity of lineage differentiation and are termed multipotential adult progenitor cells (MAPCs) (Jiang et al. 2002).

3.2   Derivation of hESCs

Human ESCs (hESCs) can be derived from the inner cell mass of the blastocysts produced in vitro for infertile couples (Trounson 2001) but may also be derived from morula stage embryos (see Fig. 1), which may demonstrate a further divergence from mouse ESCs (mESCs) in their nature and response to messengers for differentiation. For example,

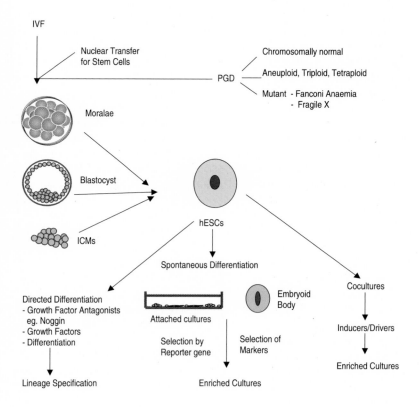

Fig. 1. Production and differentiation of hESCs

mESCs are maintained in the presence of leukaemia inhibiting factor (LIF) in culture in vitro without feeder cells, but this is not the case for hESCs (Daheron et al. 2004; Trounson 2001). It is possible to direct mESCs into a trophectoderm lineage and to establish permanent trophectoderm cell lines (Rossant 2001; Tanaka et al. 2002) but this cannot be replicated for hESCs despite the apparent expression of some trophectodermal markers in response to bone morphogenic protein 4 (BMP4) (Xu et al. 2002b).

The preliminary experiments involved in the derivation of the original hESCs by our own research group was summarised by Trounson (2001), and the methods finally used to establish hESC lines described by Reubinoff et al. (2000). The methods were similar to that described by Thomson et al. (1998) and involved the isolation of inner cell mass clusters from human blastocysts by immunosurgery and their coculture with mitotically inactivated murine embryonic fibroblasts to form typical colonies of undifferentiated cells, which need to be passaged weekly, or more often as dissected colonies of 10 cells or more. Similar methods have been used more recently to derive additional human embryonic stem cell lines (Cowan et al. 2004), and all these were based on the original methods described by Thomson and colleagues for non-human primates (Thomson et al. 1995, 1996).

It is important to note that hESCs are different from the inner cell mass (ICM) cluster. Firstly, ICM cells retain a memory for axes; dorsal–ventral, anterior–posterior and left–right axes, which enables the differentiating cells to have position relationships that guide the formation and integration required to form an organism. It is considered that ESCs are an epiblast derivative that are maintained under laboratory conditions, in the presence of LIF or secretory products of embryonic or adult somatic cells. Importantly, the self-renewal of hESCs appears to involve the Wnt family signalling pathway (Sato et al. 2004).

The criteria used for choosing embryos for deriving hESCs will determine the eventual success rates for their production. Small numbers of blastocyst stage embryos were used by Reubinoff et al. (2000) to produce six hESC lines after preliminary experiments involving around 30 embryos (Trounson 2001). The six hESC lines were derived from 12 blastocysts; in one, no ICM was identified. This very high success rate can be compared with much larger numbers of embryos/blastocysts produced by in vitro fertilisation (IVF) by others. Around 50% of human embryos have chromosomal abnormalities (Gianaroli et al. 2002), and it would be expected that this would limit the success rate of hESC production by 50% or more. It is also difficult to establish hESCs from monosomic or trisomic embryos, with less than 10% made from human embryos diagnosed as aneuploid (Verlinksy et al. 2004). Interestingly, two hESC lines produced from trisomic embryos reverted to diploidy, indicating the embryos were probably mosaic. A large number of hESC lines have been produced from excess human IVF embryos by some IVF clinics. For example Kukharenko et al. (2004) reported 46 new hESC lines made from morulae, blastocysts and ICMs (see Fig. 1) isolated from blastocysts (Strelchenko et al. 2004). There was apparently little difference between embryonic stages and the capacity to develop hESC lines.

There is a wide range of feeder cells that are appropriate for the maintenance of hESCs, including murine STO cells, murine embryonic fibroblasts (Reubinoff et al. 2000; Thomson et al. 1998) and human cell lines, including human embryonic fibroblasts (Richards et al. 2002) and human adult bone marrow cells (Cheng et al. 2003). Other cell lines, including commercially available human cell lines can be used for the maintenance of hESC and may also be appropriate for deriving the hESC from human embryos or ICMs.

3.2.1   Selection of Embryos for Deriving hESCs

The clear preference for deriving hESCs would be euploid embryos selected after embryo biopsy for preimplantation genetic diagnosis (PGD). This technique involves multicolour fluorescent in situ hybridisation (FISH). Usually one or two cells are removed from eight-cell human embryos prior to compaction (60–72 h after insemination) and 5–14 chromosome numbers are examined by multicolour fluorescent tags. Usually monosomic or trisomic embryos are discarded. Some embryos are mosaic for chromosomal numbers or are chromosomally chaotic. Very few of these embryos can develop normally in vitro or in vivo (Katz-Jaffe et al. 2004). However, hESCs that are trisomic for chromosome 13 and are triploid have been derived from human embryos, which indicates that some of the aneuploidies observed in embryos are compatible for ESC development (Heins et al. 2004).

Normally blastocyst-stage embryos (Jones 2000) are chosen for derivation of hESCs, having well-developed ICMs that are isolated mechanically or by immunosurgery (Hovatta et al. 2003; Mitalipova et al. 2003; Park et al. 2003; Reubinoff et al. 2000; Richards et al. 2002; Stojkovic et al. 2004a; Thomson et al. 1998). ICMs will form rounded cell colonies of small tightly packed cells with a large nucleus to cytoplasmic ratio (Sathananthan et al. 2001). Serum-free culture systems containing serum substitutes and fibroblast growth factor-2 (FGF-2) have now replaced serum-containing media (Amit et al. 2000) for propagation of hESCs and reduced spontaneous differentiation.

It will be important to derive hESCs from embryos produced under Good Medical Practice (GMP) and a number of IVF clinics in the United Kingdom have upgraded their facilities to GMP with the assistance of the UK Medical Research Council. Consequently, hESC lines produced under these regulations will be preferable for any clinical applications. These hESC lines will be available for research through the UK Stem Cell Bank. Similar approaches are being adopted in other countries. GMP regulations require the replacement of serum supplements and cell feeder co-cultures, particularly animal cells as feeders. Consequently, culture conditions are changing to serum-free and cell feeder-free systems. Progress towards these requirements has been reported by Xu et al. (2001), who replaced cell feeders by extracellular matrix (Matrigel) and cell feeder conditioned medium. Amit et al. (2000) showed that hESCs can be maintained in medium-containing serum replacement FGF-2, transferring growth factor-α (TGF-α) and LIF, and fibronectin extracellular matrix. It is also reported that modulation of Wnt signalling can maintain hESCs in serum- and cell feeder-free culture conditions for at least several passages (Sato et al. 2004).

Recently it has been found that hESCs can be maintained by sphingosine-1-phosphate (S1P) and platelet-derived growth factor (PGDF) in serum-free cultures. These data show that signalling pathways for hESC renewal is activated by tyrosine kinases synergistically with those downstream from lysophospholipid (LPL) receptors (Pebay et al. 2003).

It is also possible to derive hESCs from embryos with diagnosed mutations by PGD. Human ESC lines from embryos with Fanconi Anaemia-

A mutation and an expansion mutation in the FMRI gene (Fragile X mutation) (Galat et al. 2004). These hESC lines will be extremely valuable for research into the phenotypic changes induced by these mutations and provide an opportunity to develop drug treatments to reduce, or to control, these defects.

3.2.2   Genetic Manipulation of hESCs

While clonal derivation of hESCs is difficult and the efficiency is extremely low, it is possible to transfect hESCs with DNA constructs. This is important for determining the role of transcription factors for the renewal and differentiation of hESCs. Conventional transfection methods have been successful (Eiges et al. 2001), as have lentiviral methods (Gropp et al. 2003; Ma et al. 2003). Integration of reporter genes into controlling elements of specific genes or the approach of gene knock-out or knock-in, used for functional genomics, is very difficult because of the inability to clone hESCs. Zwaka and Thomson (2003) have shown that it is possible to electroporate hESCs to achieve homologous recombination of hESC colony fragments. Gene function may be more appropriately determined in hESCs by using small inhibitory RNAs (siRNA) (Vallier et al. 2004) to control renewal, differentiation, apoptosis, etc.

3.2.3   Markers of hESC Pluripotentiality

Sperger et al. (2003) have reported by microarray analysis, 330 genes that are highly expressed in hESCs and human embryonal carcinoma cells (hECCs) and seminomas. This included *POU5F1* (*Oct4*) and *FLJ10713* (a homologue highly expressed in mESCs [Ramalho-Santos et al. 2002]). Among those only highly expressed in hESCs and hECCs included a DNA methylase (*DNMT3B*), which functions in early embryogenesis (Watanabe et al. 2002) and *FOXD3*, a forkhead family transcription factor that interacts with *Oct4*, which is essential for the maintenance of mouse primitive ectoderm (Hanna et al. 2002). *SOX2* is also highly expressed and is known to be important in pluripotentiality (Avilion et al. 2003).

Markers that are now recognised as important for hESC pluripotentiality include *Oct4*, *NANOG*, *SOX2*, *FOXD3*, *REX1* and *UTF1* transcription factors; *TERF1*, *CHK2*, *DNMT3* DNA modifiers; *GFA1* surface marker; *GDF3* growth factor; *TDGF1* receptor; and *STELLA* and *FLJ10713* (Pera and Trounson 2004).

For characterisation of hESC, it is common to report one or more of the following: *Oct4* expression, alkaline phosphatase and telomerase activities; stage-specific embryonic antigens (SSEA)-3 and -4; hESC antigens TRA-1-60, TRA-1-81, GCTM-2, TG-30 and TG-343; and CD9, Thy1 and major histocompatibility complex class 1 (Stojkovic et al. 2004b).

3.3   Differentiation of hESCs

Spontaneous differentiation of hESC colonies occurs in prolonged cultures and in the absence of actively secreting feeder cells. Early differentiation events may be observed in many hESC colonies within a week of passage (Sathananthan et al. 2001) and heterogeneity in markers of stem cell pluripotentiality (e.g. Oct 4) can be observed in hESC colonies. When hESCs are permitted to overgrow in two-dimensional culture, cells begin to pile up and differentiation occurs at the borders of the colony and in the central piled up areas of the colony (Sathananthan et al. 2001). A wide range of differentiating cell types can be observed in these flat cultures, including ectodermal neuroectoderm, mesodermal muscle and endodermal organ tissue types (Reubinoff et al. 2000).

A more sophisticated culture system involves the formation of embryoid bodies (Itskovitz-Eldor et al. 2000) that involves culture of hESCs in 'hanging drops' or in plastic culture dishes that do not favour cell adhesion and attachment. Embryoid bodies have a consistent appearance and structure (Conley et al. 2004; Sathananthan 2003), with a variety of cell types that appear to develop in a more random organisation than mouse embryoid bodies. Visceral endoderm is consistently identified on the outer surface of human embryoid bodies (Conley et al. 2004; Sathananthan 2003). With a variety of cell types produced in embryoid bodies, it is possible to select specific cell populations of interest using surface markers and cell separation techniques, including fluorescent cell sorting (Levenberg et al. 2002). It is also possible to use lineage-specific promoters driving reporter genes or selective cell morphology

(Reubinoff et al. 2001). Cultures have been significantly enriched for cardiomyocytes using buoyant density gradient separation methods and marker selection (Kehat et al. 2001; Xu et al. 2002a).

3.3.1   Directing Differentiation

The enhancement of differentiation towards a specific lineage (see Fig. 1) can be achieved by activating endogenous transcription factors or transfection of ESCs with ubiquitously expressing transcription factors, by exposure of ESCs to growth factors or coculture of ESCs with inducing cells (Trounson 2005).

Coculture-Induced Differentiation

Coculture methodologies have been used to produce blood cell precursors (Kaufman et al. 2001), dopaminergic neurones (Kawasaki et al. 2002), neural crest cells and motor neurones (Mizuseki et al. 2003), cardiomyocytes (Mummery et al. 2003), and lung alveoli (Denham et al. 2002).

Mummery et al. (2002; 2003) showed that 15%–20% of cultures of hESCs grown with the mouse visceral endoderm cell type END-2, will form beating-heart muscle colonies, and this has been substantially increased in more recent experiments. Action potentials of cardiomyocytes produced in this system generally resemble that for human embryonic/fetal left ventricular cardiomyocytes, but are distinctly different from those of mouse cardiomyocytes. Atrial- and pacemaker-like cells may also be formed in the differentiating hESC cultures. The hESC-derived cardiomyocytes are capable of integrating into rodent heart muscle (C. Mummery, personal communication). This research has advanced the prospect of hESCs being used in the clinical treatment of cardiac damage.

Studies on the successful induction of mouse alveolar phenotypes from mESCs that have the morphological appearance of type II pneumocytes and that express surfactant protein C (SPC) (respiratory specific marker) by coculture with mouse embryonic mesenchyme have also been shown to be effective for differentiation of hESCs into the respiratory lineage (Denham et al. 2002). The hESC-derived cells also expressed human SPC. This is encouraging for the further research on the use of hESCs for lung engraftment.

Co-culture of monkey ESCs with the PA6 cell line that produces stromal cell-derived inducing activity (SDIA) will produce midbrain neuronal cells that are tyrosine hydrolase positive (TH$^+$) and express *nurr1* and *LMX1b* genes (Kawasaki et al. 2002; Mizuseki et al. 2003). In these differentiating cultures, pigmented retinal epithelium could also be recognised. Manipulation of culture conditions with BMP4 induces epidermogenesis or neural crest cells and dorsal-most central nervous system cells, and suppression of *sonic hedgehog* promotes motor neurone formation (Trounson 2004).

Keratinocytes can be derived from hESCs by replating embryoid bodies (Green et al. 2003). Cells expressing the transcription factor *p63* in the periphery of the secondary cultures identify the keratinocyte progenitors that produce more mature keratinocyte cell types, which produce cytokeratin 14 and basonuclin. These cells can form terminally differentiated stratifying epithelium but were not the same as keratinocyte epithelium isolated from neonatal or adult skin.

Growth Factor Induction of Differentiation

ESCs may be induced into the lineage of interest by growth factors or their antagonists (Loebel et al. 2003). In the case of ectodermal differentiation, Pera et al. (2004) showed bone morphogenic proteins (BMPs) were involved in a paracrine loop that induced hESC differentiation into cells expressing genes characteristic of extraembryonic endoderm. By blocking BMP signalling in hESCs with the antagonist noggin, we produced a neuroectodermal precursor cell type that was capable of facile conversion into neural progenitor cells when transferred to suspension culture of basal medium supplemented with fibroblast growth factor-2 (FGF-2).

Using embryoid body cultures and a cocktail of haematopoietic cytokines and BMP-4, Chadwick et al. (2003) have induced the formation of haematopoietic progenitors that could produce erythroid and myeloid derivatives. The progenitors were immunologically similar to haematopoietic progenitors of the dorsal aorta. The growth factors used were stem cell factor (SCF), interleukins-3 and -6 (IL-3, IL-6), granulocyte colony-stimulating factor (GCSF) and Flt-3 ligand. A further enhancement of erythroid colonies can be obtained with the addition of vascular endothelial growth factor-A (VEGF-A) (Cerdan et al. 2004).

3.3.2 Selecting Lineage Cell Types of Interest

When hESCs are transplanted under the mouse testis or kidney capsule, mixed cell teratomas develop with frequently very mature phenotype (Gertow et al. 2004; Reubinoff et al. 2000; Thomson et al. 1998). The number of cell types, their maturation and their organisation are much more substantial than what is observed in in vitro differentiation systems. The teratomas formed are mainly of human origin and most organised tissue areas are exclusively human, but mouse cells are observed to take on microstructures with appropriate histiotypic appearance, indicating interactions in both directions between mouse and human tissues (Gertow et al. 2004). Consequently, it appears that the microenvironment to which ESCs are exposed will induce very different responses, and in the case of interspecies transplantation, very different capabilities for cell lineage formation, maturation and organisation into primitive embryonic-like organs.

Selection of differentiating cells of the endodermal lineage has been difficult, probably because of the lack of markers of early progenitors of this lineage. There is much interest in the production of pancreatic α-Islet cells because of the potential to treat diabetes. Some cells of embryoid bodies will stain positive to insulin antibodies (Assady et al. 2001), but while they weakly express insulin-2, they do not express insulin-1 and do not stain for C-peptide (Rajagopal et al. 2003). The production of insulin-producing cells from mESCs, differentiating from neuroectoderm in the mouse (Lumelsky et al. 2001) has been difficult to repeat with hESCs. Rambhatla et al. (2003) reported differentiation of hESCs into cells expressing markers of hepatocytes by treatment of differentiating embryoid bodies with sodium butyrate or adherent hESC cultures with dimethyl sulfoxide followed by sodium butyrate.

The selection of cells with particular morphology in adherent hESC cultures differentiating in vitro may also favour endodermal populations that express markers of fetal lines (Stamp et al. 2003). These data suggest that with the appropriate markers, it will be possible to select cells capable of forming liver, gut and other endodermal tissues (S. Hawes, personal communication).

3.4 Conclusions

hESCs are now commonly derived from excess human IVF embryos resulting from the treatment of infertile patients or those undergoing PGD for chromosomal abnormality or genetic disease. This is rapidly expanding the number of hESCs available for research and will provide a very valuable resource for cell and gene therapy. Data from directed differentiation of hESC has shown that mature neurones are capable of colonising rodent brains (Reubinoff et al. 2001; Zhang et al. 2001) and fetal-like cardiomyocytes (Mummery et al. 2003) appear to integrate successfully in rodent heart tissue. The present research is concentrated in preclinical trials in animal models for the correction of degenerative disorders and injuries. It might be expected that the first clinical trials with hESC-derived cells will occur within the next 5 years and may include the use of retinal pigmented epithelium for loss of visual function (Klimanskaya et al. 2004). It is important that the application of embryonic stem cell therapies be accompanied by the reassurance of safety and efficacy expected of drug therapies.

References

Amit M, Carpenter MK, Inokuma MS, Chiu CP, Harris CP, Waknitz MA, Itskovitz-Eldor J, Thomson JA (2000) Clonally derived human embryonic stem cell lines maintain pluripotency and proliferative potential for prolonged periods of culture. Dev Biol 227:271–278

Assady S, Maor G, Amit M, Itskovitz-Eldor J, Skorecki KL, Tzukerman M (2001) Insulin production by human embryonic stem cells. Diabetes 50:1691–1697

Avilion AA, Nicolis SK, Pevny LH, Perez L, Vivian N, Lovell-Badge R (2003) Multipotent cell lineages in early mouse development depend on SOX2 function. Genes Dev 17:126–140

Cerdan C, Rouleau A, Bhatia M (2004) VEGF-A165 augments erythropoietic development from human embryonic stem cells. Blood 103:2504–2512

Chadwick K, Wang L, Li L, Menendez P, Murdoch B, Rouleau A, Bhatia M (2003) Cytokines and BMP-4 promote hematopoietic differentiation of human embryonic stem cells. Blood 102:906–915

Cheng L, Hammond H, Ye Z, Zhan X, Dravid G (2003) Human adult marrow cells support prolonged expansion of human embryonic stem cells in culture. Stem Cells 21:131–142

Conley BJ, Trounson AO, Mollard R (2004) Human embryonic stem cells form embryoid bodies containing visceral endoderm-like derivatives. Fetal Diagn Ther 19:218–223

Cowan CA, Klimanskaya I, McMahon J, Atienza J, Witmyer J, Zucker JP, Wang S, Morton CC, McMahon AP, Powers D, Melton DA (2004) Derivation of embryonic stem-cell lines from human blastocysts. N Engl J Med 350:1353–1356

Daheron L, Opitz SL, Zaehres H, Lensch WM, Andrews PW, Itskovitz-Eldor J, Daley GQ (2004) LIF/STAT3 signaling fails to maintain self-renewal of human embryonic stem cells. Stem Cells 22:770–778

Denham M, Trounson A, Mollard R (2002) Respiratory lineage differentiation of embryonic stem cells in vitro. Keystone Symposia – stem cells, Colorado, USA

Eiges R, Schuldiner M, Drukker M, Yanuka O, Itskovitz-Eldor J, Benvenisty N (2001) Establishment of human embryonic stem cell-transfected clones carrying a marker for undifferentiated cells. Curr Biol 11:514–518

Galat V, Strelchenko N, Ozen S, Sky S, Kukharenko V, Verlinsky Y (2004) Human embryonic stem cells from embryos affected by genetic diseases (abstract 115). International Society for Stem Cell Research 2nd Annual Meeting, Boston, USA pp 77

Gertow K, Wolbank S, Rozell B, Sugars R, Andang M, Parish CL, Imreh MP, Wendel M, Ahrlund-Richter L (2004) Organized development from human embryonic stem cells after injection into immunodeficient mice. Stem Cells Dev 13:421–435

Gianaroli L, Magli MC, Ferraretti AP, Tabanelli C, Trombetta C, Boudjema E (2002) The role of preimplantation diagnosis for aneuploidies. Reprod Biomed Online 4 Suppl 3:31–36

Green H, Easley K, Iuchi S (2003) Marker succession during the development of keratinocytes from cultured human embryonic stem cells. Proc Natl Acad Sci U S A 100:15625–1630

Gropp M, Itsykson P, Singer O, Ben-Hur T, Reinhartz E, Galun E, Reubinoff BE (2003) Stable genetic modification of human embryonic stem cells by lentiviral vectors. Mol Ther 7:281–287

Hanna LA, Foreman RK, Tarasenko IA, Kessler DS, Labosky PA (2002) Requirement for Foxd3 in maintaining pluripotent cells of the early mouse embryo. Genes Dev 16:2650–2661

Heins N, Englund MC, Sjoblom C, Dahl U, Tonning A, Bergh C, Lindahl A, Hanson C, Semb H (2004) Derivation, characterization, and differentiation of human embryonic stem cells. Stem Cells 22:367–376

Herzog EL, Chai L, Krause DS (2003) Plasticity of marrow-derived stem cells. Blood 102:3483–3493

Hovatta O, Mikkola M, Gertow K, Stromberg AM, Inzunza J, Hreinsson J, Rozell B, Blennow E, Andang M, Ahrlund-Richter L (2003) A culture system using human foreskin fibroblasts as feeder cells allows production of human embryonic stem cells. Hum Reprod 18:1404–1409

Itskovitz-Eldor J, Schuldiner M, Karsenti D, Eden A, Yanuka O, Amit M, Soreq H, Benvenisty N (2000) Differentiation of human embryonic stem cells into embryoid bodies compromising the three embryonic germ layers. Mol Med 6:88–95

Jiang Y, Jahagirdar BN, Reinhardt RL, Schwartz RE, Keene CD, Ortiz-Gonzalez XR, Reyes M, Lenvik T, Lund T, Blackstad M, Du J, Aldrich S, Lisberg A, Low WC, Largaespada DA, Verfaillie CM (2002) Pluripotency of mesenchymal stem cells derived from adult marrow. Nature 418:41–49

Jones GM (2000) Growth and viability of human blastocysts in vitro. Reprod Med Rev 8:241–287

Katz-Jaffe MG, Trounson AO, Cram DS (2004) Mitotic errors in chromosome 21 of human preimplantation embryos are associated with non-viability. Mol Hum Reprod 10:143–147

Kaufman DS, Hanson ET, Lewis RL, Auerbach R, Thomson JA (2001) Hematopoietic colony-forming cells derived from human embryonic stem cells. Proc Natl Acad Sci U S A 98:10716–1021

Kawasaki H, Suemori H, Mizuseki K, Watanabe K, Urano F, Ichinose H, Haruta M, Takahashi M, Yoshikawa K, Nishikawa S, Nakatsuji N, Sasai Y (2002) Generation of dopaminergic neurons and pigmented epithelia from primate ES cells by stromal cell-derived inducing activity. Proc Natl Acad Sci U S A 99:1580–1585

Kehat I, Kenyagin-Karsenti D, Snir M, Segev H, Amit M, Gepstein A, Livne E, Binah O, Itskovitz-Eldor J, Gepstein L (2001) Human embryonic stem cells can differentiate into myocytes with structural and functional properties of cardiomyocytes. J Clin Invest 108:407–414

Klimanskaya I, Hipp J, Rezai KA, West M, Atala A, Lanza R (2004) Derivation and comparative assessment of retinal pigment epithelium from human embryonic stem cells using transcriptomics. Cloning Stem Cells 6:217–245

Kukharenko V, Strelchenko N, Galat V, Sky O (2004) Panel of human embryonic stem cell lines (abstract 143). International Society for Stem Cell Research 2nd Annual Meeting, Boston, USA pp 87

Levenberg S, Golub JS, Amit M, Itskovitz-Eldor J, Langer R (2002) Endothelial cells derived from human embryonic stem cells. Proc Natl Acad Sci U S A 99:4391–4396

Loebel DA, Watson CM, De Young RA, Tam PP (2003) Lineage choice and differentiation in mouse embryos and embryonic stem cells. Dev Biol 264:1–14

Lumelsky N, Blondel O, Laeng P, Velasco I, Ravin R, McKay R (2001) Differentiation of embryonic stem cells to insulin-secreting structures similar to pancreatic islets. Science 292:1389–1394

Ma Y, Ramezani A, Lewis R, Hawley RG, Thomson JA (2003) High-level sustained transgene expression in human embryonic stem cells using lentiviral vectors. Stem Cells 21:111–117

Mitalipova M, Calhoun J, Shin S, Wininger D, Schulz T, Noggle S, Venable A, Lyons I, Robins A, Stice S (2003) Human embryonic stem cell lines derived from discarded embryos. Stem Cells 21:521–526

Mizuseki K, Sakamoto T, Watanabe K, Muguruma K, Ikeya M, Nishiyama A, Arakawa A, Suemori H, Nakatsuji N, Kawasaki H, Murakami F, Sasai Y (2003) Generation of neural crest-derived peripheral neurons and floor plate cells from mouse and primate embryonic stem cells. Proc Natl Acad Sci U S A 100:5828–5833

Mummery C, Ward D, van den Brink CE, Bird SD, Doevendans PA, Opthof T, Brutel de la Riviere A, Tertoolen L, van der Heyden M, Pera M (2002) Cardiomyocyte differentiation of mouse and human embryonic stem cells. J Anat 200:233–242

Mummery C, Ward-van Oostwaard D, Doevendans P, Spijker R, van den Brink S, Hassink R, van der Heyden M, Opthof T, Pera M, de la Riviere AB, Passier R, Tertoolen L (2003) Differentiation of human embryonic stem cells to cardiomyocytes: role of coculture with visceral endoderm-like cells. Circulation 107:2733–2740

Park JH, Kim SJ, Oh EJ, Moon SY, Roh SI, Kim CG, Yoon HS (2003) Establishment and maintenance of human embryonic stem cells on STO, a permanently growing cell line. Biol Reprod 69:2007–2014

Pebay A, Wong R, Pitson S, Peh G, Koh K, Tellis I, Nguyen L, Pera M (2003) Maintenance of human embryonic stem cells by sphingosine-1-phosphate and platelet-derived growth factor in a serum-free medium. First National Stem Cell Centre Scientific Conference – Stem Cells and Tissue Repair 2003, pp 104

Pera MF, Trounson AO (2004) Human embryonic stem cells: prospects for development. Development 131:5515–5525

Pera MF, Reubinoff B, Trounson A (2000) Human embryonic stem cells. J Cell Sci 113:5–10

Pera MF, Andrade J, Houssami S, Reubinoff B, Trounson A, Stanley EG, Ward-van Oostwaard D, Mummery C (2004) Regulation of human embryonic stem cell differentiation by BMP-2 and its antagonist noggin. J Cell Sci 117:1269–1280

Rajagopal J, Anderson WJ, Kume S, Martinez OI, Melton DA (2003) Insulin staining of ES cell progeny from insulin uptake. Science 299:363

Ramalho-Santos M, Yoon S, Matsuzaki Y, Mulligan RC, Melton DA (2002) "Stemness": transcriptional profiling of embryonic and adult stem cells. Science 298:597–600

Rambhatla L, Chiu CP, Kundu P, Peng Y, Carpenter MK (2003) Generation of hepatocyte-like cells from human embryonic stem cells. Cell Transplant 12:1–11

Reubinoff BE, Pera MF, Fong CY, Trounson A, Bongso A (2000) Embryonic stem cell lines from human blastocysts: somatic differentiation in vitro. Nat Biotechnol 18:399–404

Reubinoff BE, Itsykson P, Turetsky T, Pera MF, Reinhartz E, Itzik A, Ben-Hur T (2001) Neural progenitors from human embryonic stem cells. Nat Biotechnol 19:1134–1140

Richards M, Fong CY, Chan WK, Wong PC, Bongso A (2002) Human feeders support prolonged undifferentiated growth of human inner cell masses and embryonic stem cells. Nat Biotechnol 20:933–936

Rossant J (2001) Stem cells from the Mammalian blastocyst. Stem Cells 19:477–482

Sathananthan AH (2003) Origins of human embryonic stem cells and their spontaneous differentiation. First National Stem Cell Centre Scientific Conference – Stem Cells and Tissue Repair, Melbourne, Australia 2003, pp 225

Sathananthan H, Pera M, Trounson A (2001) The fine structure of human embryonic stem cells. Reprod Biomed Online 4:56–61

Sato N, Meijer L, Skaltsounis L, Greengard P, Brivanlou AH (2004) Maintenance of pluripotency in human and mouse embryonic stem cells through activation of Wnt signaling by a pharmacological GSK-3-specific inhibitor. Nat Med 10:55–63

Shamblott MJ, Axelman J, Wang S, Bugg EM, Littlefield JW, Donovan PJ, Blumenthal PD, Huggins GR, Gearhart JD (1998) Derivation of pluripotent stem cells from cultured human primordial germ cells. Proc Natl Acad Sci U S A 95:13726–13731

Sperger JM, Chen X, Draper JS, Antosiewicz JE, Chon CH, Jones SB, Brooks JD, Andrews PW, Brown PO, Thomson JA (2003) Gene expression patterns in human embryonic stem cells and human pluripotent germ cell tumors. Proc Natl Acad Sci U S A 100:13350–13355

Stamp LA, Crosby HA, Hawes SM, Strain AJ, Pera MF (2003) Characterisation of GCTM5, a putative marker for early liver cells. First National Stem Cell Centre Annual Conference – Stem Cells and Tissue Repair, Melbourne, Australia 2003, pp 28

Stojkovic M, Lako M, Stojkovic P, Stewart R, Przyborski S, Armstrong L, Evans J, Herbert M, Hyslop L, Ahmad S, Murdoch A, Strachan T (2004a) Derivation of human embryonic stem cells from day-8 blastocysts recovered after three-step in vitro culture. Stem Cells 22:790–797

Stojkovic M, Lako M, Strachan T, Murdoch A (2004b) Derivation, growth and applications of human embryonic stem cells. Reproduction 128: 259–267

Strelchenko N, Kukharenko V, Verlinksy O (2004) Human ES-cells derived from different embryo stages (abstract 359). International Society for Stem Cell Research 2nd Annual Meeting 2004, pp 163

Tanaka TS, Kunath T, Kimber WL, Jaradat SA, Stagg CA, Usuda M, Yokota T, Niwa H, Rossant J, Ko MS (2002) Gene expression profiling of embryo-derived stem cells reveals candidate genes associated with pluripotency and lineage specificity. Genome Res 12:1921–1928

Thomson JA, Kalishman J, Golos TG, Durning M, Harris CP, Becker RA, Hearn JP (1995) Isolation of a primate embryonic stem cell line. Proc Natl Acad Sci U S A 92:7844–7848

Thomson JA, Kalishman J, Golos TG, Durning M, Harris CP, Hearn JP (1996) Pluripotent cell lines derived from common marmoset (Callithrix jacchus) blastocysts. Biol Reprod 55:254–259

Thomson JA, Itskovitz-Eldor J, Shapiro SS, Waknitz MA, Swiergiel JJ, Marshall VS, Jones JM (1998) Embryonic stem cell lines derived from human blastocysts. Science 282:1145–1147

Trounson AO (2001) The derivation and potential use of human embryonic stem cells. Reprod Fertil Dev 13:523–532

Trounson A (2005) Derivation characteristics and perspectives for mammalian pluripotential stem cells. Reprod Fertil Dev 17:135–141

Trounson A (2004) Stem cells, plasticity and cancer – uncomfortable bed fellows. Development 131:2763–2768

Vallier L, Rugg-Gunn PJ, Bouhon IA, Andersson FK, Sadler AJ, Pedersen RA (2004) Enhancing and diminishing gene function in human embryonic stem cells. Stem Cells 22:2–11

Verlinksy Y, Strelchenko N, Kukharenko V, Galat V (2004) Preimplantation genetic diagnosis: as a source of human embryonic stem cell lines (abstract 370). International Society for Stem Cell Research 2nd Annual Meeting., Boston, USA 2004, pp 166

Watanabe D, Suetake I, Tada T, Tajima S (2002) Stage- and cell-specific expression of Dnmt3a and Dnmt3b during embryogenesis. Mech Dev 118:187–190

Xu C, Inokuma MS, Denham J, Golds K, Kundu P, Gold JD, Carpenter MK (2001) Feeder-free growth of undifferentiated human embryonic stem cells. Nat Biotechnol 19:971–974

Xu C, Police S, Rao N, Carpenter MK (2002a) Characterization and enrichment of cardiomyocytes derived from human embryonic stem cells. Circ Res 91:501–508

Xu RH, Chen X, Li DS, Li R, Addicks GC, Glennon C, Zwaka TP, Thomson JA (2002b) BMP4 initiates human embryonic stem cell differentiation to trophoblast. Nat Biotechnol 20:1261–1264

Zhang SC, Wernig M, Duncan ID, Brustle O, Thomson JA (2001) In vitro differentiation of transplantable neural precursors from human embryonic stem cells. Nat Biotechnol 19:1129–1133

Zwaka TP, Thomson JA (2003) Homologous recombination in human embryonic stem cells. Nat Biotechnol 21:319–321

4 In Search of the Elusive Epidermal Stem Cell

R. Ghadially

Abstract. Recent studies are beginning to reveal that our basic concepts of epidermal stem cell biology may be based on somewhat tenuous ground. For example, it is often assumed that colony-forming cells represent epidermal stem cells, although this has not proved to be the case in hematopoietic cell lineages. In addition, although most stem cells are not cycling, label-retaining cells are used as a primary measure of epidermal stem cells. Moreover, the locations of stem cell niches in epidermis are still being debated. Finally, while putative stem cell markers abound, the most effective isolation procedure for stem cells has not been determined, and the relative efficiency of various methods of stem cell isolation remains unknown. With a functional assay for epidermal stem cells (analogous to the in-vivo competitive assay used for hematopoiesis), we appear to be in a better position to more clearly define the molecular signature of the true long-term repopulating cell/stem cell of the epidermis. Nonetheless,

significant progress has been made in regenerative therapy of the epidermis
for ulcer and burn treatment, and for corrective gene therapy for inherited skin
diseases

4.1   Introduction

4.1.1   The Epidermal Proliferation Unit

Integrity of the epidermis is maintained by division of cells in the proliferative basal layer to replace differentiated cells in the outermost stratum
corneum layer that are continuously being lost. Stem cells in the basal
layer divide to produce, on average, one stem cell and one transit amplifying cell (Lavker and Sun 2000). The transit amplifying cells amplify
the basal cell population, but are limited to a finite number of cell divisions before they differentiate and are sloughed into the environment
(Potten 1981). Stem cells are thought to constitute 1%–10% of the basal
cell population (Mackenzie and Bickenbach 1985; Morris et al. 1985;
Bickenbach et al. 1986; Morris and Potten 1994; Heenen and Galand
1997).

The epidermis is composed of discrete epidermal proliferation units
(Potten 1981). In mice, each unit lies over approximately ten basal
cells (the central one being the stem cell), thought to be responsible for
the generation and maintenance of the superficial column (composed
of less than 50 cells) (Potten 1981). However, in many studies much
larger units of cells, presumably originating from one stem cell have
been observed (Schmidt et al. 1987; Mackenzie 1997; Ghazizadeh and
Taichman 2001; Zhang et al. 2001; Schneider et al. 2003). Furthermore,
several lineage analysis experiments have provided evidence that one
stem cell supplies more than one epidermal proliferation unit (Kameda
et al. 2003; Ro and Rannala 2004). Thus, it seems likely that epidermal proliferation units are founded by early progenitors derived from
a rarer stem cell. The early progenitors would give rise to the transit amplifying cells, which divide to produce terminally differentiated
cells.

4.1.2 Epidermal Stem Cell Niches

There is considerable evidence for a long-term repopulating cell in the interfollicular epidermis (Mackenzie and Bickenbach 1985; Morris et al. 1985; Ghazizadeh and Taichman 2001; Niemann and Watt 2002), including our recent publication (Schneider et al. 2003). However, debate remains over whether the interfollicular stem cell is a true stem cell or an early progenitor derived from the follicular stem cell (Mackenzie and Bickenbach 1985; Morris et al. 1985; Ghazizadeh and Taichman 2001; Niemann and Watt 2002; Cotsarelis et al. 1999; Taylor et al. 2000; Oshima et al. 2001; Fuchs and Raghavan 2002). It is thought by many that cutaneous epithelia contain stem cells from a common ectodermal origin that are equipotent, albeit located in different niches (Cotsarelis et al. 1999; Niemann and Watt 2002). One prevailing model of epithelial stem cell function is that, under steady state conditions, stem cells are unipotent (i.e., producing only follicular or only interfollicular epidermis),whereas during regeneration after tissue damage, stem cells display multipotency (reviewed in Slack 2000) (Fig. 1a). Recent lineage studies suggested that follicular stem cells provide keratinocytes in the immediate follicular rim but that interfollicular epidermal proliferation

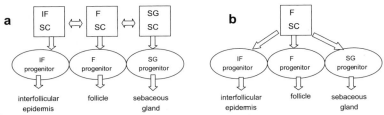

Fig. 1. Under hypothesis **a**, the interfollicular, follicular, and sebaceous gland epidermal stem cells are equipotent and multipotent. Under hypothesis **b**, the follicle is the only source of the epidermal stem cell, while interfollicular epidermis and sebaceous glands contain early progenitor cells, an intermediate-term repopulating cell, intermediate between a true stem cell and a transit amplifying cell, replaced when necessary from the follicular epidermal stem cells. These hypotheses are somewhat simplified, choosing, for example, to refrain from examining the controversy over the number and location of stem cell pools in the follicle. *SC*, stem cell; *F*, follicular; *IF*, interfollicular, *SG*, sebaceous gland

units maintained the interfollicular epidermis over 9 months in vivo (Ghazizadeh and Taichman 2001). These studies provide strong evidence for an independent stem cell in the interfollicular epidermis.

Others propose that the epidermal stem cells that lie in the hair follicle bulge are ultimately responsible for replenishing the interfollicular epidermis and sebaceous glands in addition to the hair lineages (Taylor et al. 2000; Oshima et al. 2001) (Fig. 1b). Indeed, there is considerable evidence for stem cells located in the follicular bulge region, and further evidence that follicular cells repopulate the epidermis when the epidermis is wounded (Lavker and Sun 2000; Taylor et al. 2000; Tumbar et al. 2004).

Some evidence pertains to the multipotency exhibited by 'interfollicular' stem cells. There is evidence that hairless skin possesses pluripotent stem cells (Lavker and Sun 1983), which, under the appropriate conditions, can produce hair follicles. Even though there are no hair follicles in the palms and soles, nor in foreskin (the source of human keratinocytes for most in vitro studies), implantation of dermal papillae results in hair development in these nonhairy sites, (Reynolds and Jahoda 1992; Jahoda et al. 1993). It should be noted however, that these studies investigate whether nonhairy skin, not interfollicular skin, is multipotent. This is a somewhat different question since obviously the nonhairy skin contains a primitive stem cell, while perhaps (if the most primitive pluripotent epidermal stem cell in hairy epidermis is restricted to the hair follicle) the interfollicular epidermis does not. Thus while nonhairy skin is multipotent, interfollicular epidermis may or may not be. This issue is germane not only for stem cell isolation but also for gene therapy.

4.2   Studying Epidermal Stem Cells

4.2.1   Colony-Forming Assays

Phenotypic analysis of hematopoietic stem cells has provided the ability to separate the long-term proliferating cell from the cells detected in colony forming assays (Hodgson and Bradley 1979; Van Zant 1984; Lerner and Harrison 1990; Spangrude and Johnson 1990; Morrison and Weissman 1994; Trevisan and Iscove 1995; Randall et al. 1996;

Weissman 2002). These studies and others have allowed a hierarchy of hematopoietic stem cell differentiation to be determined (for hierarchical map of phenotypes see Terskikh et al. 2003, Fig. 1). Moreover, it should be noted that populations that represent the closest stem cell descendants that can be prospectively isolated from the true stem cell by flow cytometry, are still oligopotent, yet already have a decline in self-renewal capacity, underscoring the sine qua non of a stem cell as its long-term repopulating ability (Kondo et al. 1997; Akashi et al. 2000).

The colony-forming efficiency of freshly isolated keratinocytes (in defined medium, on irradiated 3T3 fibroblasts) from human epidermis has been studied. Colony-forming efficiency of freshly isolated neonatal human keratinocyte basal cells was 0.09% or 1 in 1,000 basal cells (Wilke et al. 1988). In this study, the colony-forming efficiency of suprabasal cells was greater than that of basal cells: 0.43% of the suprabasal cells and 0.36% of the upper epidermal fraction were able to form colonies (Wilke et al. 1988). Some of the colony-forming cells were positive for involucrin. Thus it appears likely that a large number of cells that are not basal cells, and thus not likely to be stem cells, are capable of forming colonies.

In vitro studies have shown that the cultured keratinocytes most rapidly adherent to collagen IV are most enriched in colony forming cells (Jones and Watt 1993). Whether this population of rapidly adherent cells includes a greater proportion of long-term repopulating cells than the slowly adherent cells was not studied. There is evidence that transit amplifying cells as well as some postmitotic cells retain their adhesive properties and are part of the rapidly adhesive population (Kaur and Li 2000).

Whether colony-forming units reflect true long-term functional capabilities in the epidermis remains unresolved. However, it seems likely that the long-term repopulating stem cell is either a subset of, or a different set of cells from the colony-forming cell. Thus colony-forming cells in short-term cultures may represent intermediate progenitors rather than true epidermal stem cells, because of their relatively high frequency and their high level of cell cycling.

4.2.2   Long-Term Culture

Previous efforts to study human epidermal stem cells have been largely based on studies in short-term in vitro culture. In an effort to identify cells with a high proliferative capacity in the epidermis, Barrandon and Green (Barrandon and Green 1987) have studied the colony-forming cells of the epidermis in replating assays. In these studies, greater than 50% of cultured cells formed colonies. For clonal analysis of a population of keratinocytes 50–100 cells were individually plated. At 6 days, the colonies that are produced were resuspended and each plated into a 100-mm Petri dish. Different cell types found different types of colonies in culture over the next 12 days (Barrandon and Green 1987). Cells that form mostly large smooth colonies with less than 5% of small abortive colonies are termed holoclones. Paraclones are terminally differentiating cells that form small and abortive colonies. Cells intermediate between stem and differentiated cells form meroclones, are intermediate in appearance and have reproductive capacity. In the studies of Barrandon and Green (Barrandon and Green 1987), cultured keratinocytes produced 28% holoclones, 49% meroclones, and 23% paraclones. It is believed that the holoclones are stem cells (Barrandon and Green 1987; Barrandon et al. 1988; Jones and Watt 1993) and that the paraclones are transit amplifying cells, thereby suggesting a hierarchy of epidermal stem cells beginning with an epidermal stem cell that gives rise to a continuum of cell populations with progressively diminishing capacity to proliferate and self-renew.

4.2.3   Label-Retaining Cells

While human studies of stem cells are restricted to in vitro colony studies, studies in mice have utilized the concept that a label-retaining cell is a stem cell. This method utilizes tritiated thymidine or 5-bromo-2′-deoxyuridine (BrdU) to generate label-retaining cells. Injections are given repeatedly at a time when the tissue is hyperproliferative so that all dividing cells are labeled. Weeks to months later, the cells that rarely divide retain label and are considered stem cells (Bickenbach 1981). Issues with this method include the concern that since very few stem cells are dividing at any time, only a small subset of stem cells would be identified. Furthermore, after damage caused by incorporation of a nucleotide analog, the cells may not be capable of dividing. This however, does not appear to be the case, as label-retaining cells have been stimulated to divide in recent studies (Morris and Potten 1994).

4.2.4   Long-Term Repopulation In Vivo

Despite decades of work to characterize the epidermal stem cell there still is no standardized quantitative assay to measure it. We designed a quantitative in vivo assay based on well-validated functional assays for hematopoietic stem cells (Harrison 1980; Chang et al. 2000). Using competitive long-term repopulating assays combined with a limiting dilution design, stem cell frequency in the bone marrow has been determined to be 1 in 10,000 (Szilvassy et al. 1990; Taswell 1981). This type of assay has become a standard in hematopoiesis for the analysis of stem cell markers (Spangrude et al. 1988; Weissman 2002) and for studies of stem cell differentiation and regulation (Smith et al. 1991; Mikkola et al. 2003; Park et al. 2003), and in studying other populations of stem cells quantitatively (e.g., aging/disease/cancer) (Lapidot et al. 1994; Chen et al. 1999; Sudo et al. 2000; Dick 2003).

In competitive repopulation assays, one population, the competitor, serves as a standard for repopulating potential. The second population, the test population (which possesses a marker to distinguish it from the competitor cells; in our case GFP expression), is of unknown stem cell content, and is measured relative to the populating ability of the competitor population. Long-term repopulation is required so that any marked test cells that are not stem cells will have differentiated and been lost from the epidermis, such that remaining marked test cells represent stem cells originally added and their progeny. Our assay for epidermal stem cells exploits the ability of dissociated keratinocytes to reform a cornified stratified epithelium when implanted atop the subcutaneous fascia of immunodeficient mice (Fusenig et al. 1978; Wang et al. 2000). In a limiting dilution analysis, a constant number of competitor cells and progressively fewer test cells are used. Eventually, dilutions of test cells are reached that result in no, or only very rare, recipients showing any marked cells after long-term repopulation. At 9 weeks, resultant epidermis is graded as positive or negative for GFP-positive stem cells and their progeny (competitive repopulating units) by whole mount microscopy. Using statistical methods developed for the analysis of experimental data from limiting dilution studies, based on the Poisson probability equation, the number of stem cells in the test population is calculated with the help of computer software developed for this specific application (L-Calc, Stemsoft Inc., www.stemcell.com, and/or GEMOD procedure in SAS, Cary, NC). At limiting dilution, 62.5% of the recipients have detectable marked cells.

By studying fluorescent keratinocytes in the competitive repopulating assay at limiting dilutions, we have measured the frequency of functional epidermal stem cells in murine interfollicular epidermis. On the order of 0.01% of basal cells were long-term repopulating cells, a frequency much lower than previous estimates (Schneider et al. 2003). It is notable that the frequency of epidermal stem cells in the basal layer determined by this type of assay (in the order of 1 in 10,000) was similar to the frequency of hematopoietic stem cells in the marrow (0.01%) (Taswell 1981; Szilvassy et al. 1990).

Previous estimates of epidermal stem cell frequency, based on quantitative radiobiological experiments are in the order of 10% of basal cells (Potten and Hendry 1973; Mackenzie and Bickenbach 1985; Morris et al. 1985; Bickenbach et al. 1986; Morris and Potten 1994; Heenen and Galand 1997, reviewed in Cotsarelis et al. 1999), while 50% are thought to be transit amplifying cells and 40% postmitotic cells (Heenen and Galand 1997). However, in murine skin only 0.2% of basal cells retain large amounts of label without significant loss over 30–72 days (Bickenbach 1981). Also, colony forming efficiency was 45 ± 8.5 per 10,000 cells or 5 in 1,000 murine epidermal cells (Morris et al. 1988). Since just under 50% of murine epidermal cells are basal cells (Schneider et al. 2003) this means that 5 in 500, or in the order of 1 in 100, basal cells were clonogenic. It is difficult to reconcile this with the estimate of 1 in 10 basal cells being a stem cell, if stem cells are indeed clonogenic. Finally, in hematopoiesis colony-forming units, originally thought to represent stem cells, were shown to be intermediate-term repopulating cells, while true long-term repopulating stem cells were much less frequent (Chang et al. 2000). Thus further studies using the long-term repopulating assay in vivo will aid in determining the relationship between colony-forming cells and long-term repopulating cells of the epidermis.

4.3 Identifying Enriched Populations of Epidermal Stem Cells

A number of strategies have been proposed to enrich epidermal stem cells, but controversy remains regarding the best epidermal stem cell marker (Alonso and Fuchs 2003) and the degree of the enrichment provided by each method is unknown. Since functional assays to identify stem cells have been lacking for epidermal stem cells, cell surface integrin expression was used to identify a pool of $\beta 1$ integrin high cells that exhibited high colony formation and that were classified as stem cells in comparison to a more proliferative transit amplifying cell pool (Jones and Watt 1993). The use of cell-surface expression markers to isolate putative stem cells by FACS also led to claims of identification and isolation of keratinocyte stem cells using $\alpha 6$ integrin bright and transferrin receptor dim as markers (Kaur and Li 2000; Tani et al. 2000). These investigators found that their "stem cell" pool met the criteria of being label-retaining, quiescent, high nuclear to cytoplasmic ratio, high colony formation capacity, and the cells within the bulge were transferrin dim. They estimated that their "stem cell" pool represented 8% of the basal cell layer, which was consistent with that proposed by earlier studies (Potten and Morris 1988; Mackenzie 1997). Interestingly, in recent studies transcriptional profiles revealed upregulation of $\alpha 6$ and $\beta 1$ integrins in hematopoietic, neural, and embryonic stem cells (Ivanova et al. 2002; Ramalho-Santos et al. 2002). However, in a new finding, this laboratory recently reported that the $\alpha 6$ integrin bright and transferrin receptor bright pool that they had identified as the transit amplifying cells, as well as the $\alpha 6$ integrin dim pool that they identified as more committed basal cell progenitors, were almost equally as efficient as their "stem cells" in the ability to form a complete epidermis in both lifted cultures and in an in vivo assay, given a supporting dermal equivalent (Li et al. 2004). These authors thus cautioned against the use of most of the commonly accepted criteria for the identification of stem cells. These findings complement those of Wilke et al. who showed that most of the proliferating cells in culture were derived from suprabasal cells (Wilke et al. 1988). Taken together with data suggesting that stem cells represent 10% of the basal cell layer and recent studies (discussed above) suggesting that each stem cell supplies more than one epidermal proliferation unit, the previously isolated "stem cells" may be more akin to the CD34+ pool of hematopoietic progenitors, 1% of which are the functional stem cells.

Properties of adhesion and low proliferation typify stem cells and thus combining an adhesion marker with a proliferation/differentiation marker appears to be an attractive approach. In the last year, multiple proliferation/differentiation markers have been reported. CD71, the transferrin receptor, was discussed above. Cells that adhere to collagen I within 12 min and have a low cell-surface expression level of the epidermal growth factor receptor displayed increased total cell output in culture (Fortunel et al. 2003). Desmoglein is a constitutive protein of desmosomes. Desmosomal protein expression increases as keratinocytes mature. It was shown that human keratinocytes with low expression of desmoglein have high clonogenicity, colony-forming efficiency, and enhanced proliferative potential (Wan et al. 2003). Connexin 43 is a gap junction protein present in the basal layer of normal human epidermis. A recent study showed that about 10% of the basal keratinocytes were connexin 43 negative, as determined by flow cytometry. The cells were small and low in granularity, and most label-retaining cells did not express connexin 43, suggesting that this may be a negative marker of epidermal stem cells (Matic et al. 2002; Matic and Simon 2003).

Side population cells are a small population of cells that were first identified in hematopoietic cells on the basis of their Hoescht 33342 efflux (Goodell et al. 1996). Side population cells in bone marrow are enriched at least 1,000-fold in long-term hematopoietic repopulation assays. In tissues such as brain, pancreas, and muscle, side population cells may also represent stem cells (Gussoni et al. 1999; Hulspas and Quesenberry 2000; Lechner et al. 2002). Attempts have been made to isolate both murine and human epidermal side population cells. Using a modification of the original method of Goodell, a Hoechst and propidium iodide dye combination and specifically defined gating, investigators showed that mouse epidermal basal cells could be sorted into three fractions. More than 90% of the putative stem cells showed a G0/G1 cell cycle profile. Stem cells formed larger, more expandable colonies than the other fractions, in culture (Dunnwald et al. 2001). Murine side population cells were also isolated by Zhou et al. (2004). They were found to express keratin 14, $\beta1$ integrin, and p63. More recently, human epidermal side population cells were isolated. This population was enriched in quiescent cells. However, it was a distinct population from the label-retaining cell population and had low expression of surface antigens traditionally thought to mark stem cells (Terunuma et al. 2003). Finally, another group also isolated both murine and human side population cells and found that they represented a subset of the $\alpha6$ integrin-positive cells. Human side population cells did not express high levels of $\beta1$ integrin. However, they expressed the drug transporter ABCG2 (Triel et al. 2004). Because side population cells can be isolated from epidermis, comprise a small number of cells (Triel et al. 2004), do not express the markers associated with most proliferative cells, and have been shown to isolate stem cells in multiple tissues, this is a candidate marker for one that might somewhat specifically, if not uniquely, identify the true long-term repopulating stem cell. However, only in hematopoiesis have side population cells been shown to represent the true long-term repopulating cells in vivo.

Many studies have focused on the hair follicle bulge region as the site of epidermal stem cells. Keratin 19 has been shown to co-localize with [3][H] Thymidine retaining follicular bulge keratinocytes in mice. Keratin 19 did not label [3][H] Thymidine retaining cells of the interfollicular epidermis of hairy sites (Michel et al. 1996). Keratin 15 in the follicular bulge was shown to co-label with label-retaining cells and with a high level of $\beta1$ integrin expression (Lyle et al. 1998). Recently a cell surface marker, CD34, was reported as a specific marker of bulge keratinocytes. CD34 positive cells were predominantly in G0/G1, and had higher $\alpha6$ integrin expression than CD34– cells (Trempus et al. 2003). However, in a prior study label-retaining cells did not express CD34 (Albert et al. 2001).

In a recent study of cultured human keratinocytes, the cells of holo-clones (in vitro clones that show less than 5% terminal colonies) showed high expression of p63 (Pellegrini et al. 2001). However, whether the cells that show high expression of p63 display the characteristics of stem cells was not determined.

The results from these various studies are difficult to compare as different populations of cells (human, murine, follicular, nonfollicular) were studied using different methods of analyzing "stemness." Several of the methods test "stemness" by the formation of colony forming units over 1–2 weeks, a method that has not yet been shown to reflect the true long-term repopulating cell of the epidermis. Measurements of short-term colony-forming units (CFUs) clearly do not correlate with long-term functional capabilities in the blood. A detailed account of disadvantages of CFU studies is given in (Harrison 1980). Thus, whether or not CFUs reflect true long-term functional capabilities in the epidermis is not known, but seems unlikely. Only quantifying the degree to which functional stem cells are isolated by these different methods will allow the molecular signature of the epidermal stem cell to be determined, and thus provide us with the best approach(es) for enriching and/or isolating populations of epidermal stem cells. Based on findings in the hematopoietic system, it is anticipated that a combination of positive and negative surface markers will be required to identify epidermal stem cell and that the negative markers will be as key as the positive markers.

4.4   Toward Regenerative Medicine for the Epidermis

Although we lack a complete understanding of much epidermal stem cell biology, we already graft both autologous and allogenic keratinocytes for healing of burns and ulcers. Autologous keratinocytes are easy to obtain, able to be cultured ex vivo, and immunologically optimal. The downsides are the second site defect and the time required to produce the ex vivo epidermis (3–4 weeks). Successful transplantation of cultured preconfluent keratinocytes has been performed using multiple different delivery methods, making a 3- to 4-week wait unnecessary (Chester et al. 2004). Keratinocytes have been applied directly to the wound. Membrane delivery systems are also available, on which keratinocytes are cultured and then peeled off for use, and include fibrin glue, collagen 1, and hyaluronic acid. Synthetic polymers such as polyurethane and Teflon film have also been used (Chester et al. 2004). Although there are a multiplicity of systems described, large substantial clinical trials are still needed. Ideally, rather than culturing autologous cells in vitro, it would be advantageous to expand the pool of epidermal stem cells ex vivo and return an expanded pool of stem cells to the wound. This would be a pool of cells with the greatest proliferative potential for the long-term.

Finally, keratinocyte stem cells are an attractive target for gene therapy of genetic disorders of the skin as well as a possible vehicle for delivery of substances internally to correct systemic genetic defects. In the first category are diseases such as xeroderma pigmentosum (Magnaldo and Sarasin 2002, 2004), ichthyoses such as lamellar ichthyosis (Choate et al. 1996) and X-linked ichthyosis (Freiberg et al. 1997), and mechanobullous disorders such as some forms of epidermolysis bullosa (Dellambra et al. 1998). Isolation of stem cells, and determining their location(s) will help advance these fields

References

Akashi K, Traver D, Miyamoto T et al (2000) A clonogenic common myeloid progenitor that gives rise to all myeloid lineages. Nature 404:193–197

Albert MR, RA Foster, Vogel JC et al (2001) Murine epidermal label-retaining cells isolated by flow cytometry do not express the stem cell markers CD34, Sca-1, or Flk-1. J Invest Dermatol 117:943–948

Alonso L, Fuchs E (2003) Stem cells of the skin epithelium. Proc Natl Acad Sci U S A 100 Suppl 1:11830–11835

Barrandon Y, Green H (1987) Three clonal types of keratinocyte with different capacities for multiplication. Proc Natl Acad Sci U S A 84:2302–2306

Barrandon Y, Li V, Green H et al (1988) New techniques for the grafting of cultured human epidermal cells onto athymic animals. J Invest Dermatol 91:315–318

Bickenbach JR (1981) Identification and behavior of label-retaining cells in oral mucosa and skin. J Dent Res 60 [Spec No C]:1611–1620

Bickenbach JR, McCutecheon J, Mackenzie IC (1986) Rate of loss of tritiated thymidine label in basal cells in mouse epithelial tissues. Cell Tissue Kinet 19:325–333

Chang H, Jensen LA, Queensberry P (2000) Standardization of hematopoietic stem cell assays: a summary of a workshop and working group meeting sponsored by the National Heart Lung, and Blood Institute held at the National Institutes of Health Bethesda MD on September 8–9, 1998 and July 30, 1999. Exp Hematol 28:743–752

Chen J, Astle CM, Harrison DE (1999) Development and aging of primitive hematopoietic stem cells in BALB/cBy mice. Exp Hematol 27:928–935

Chester DL, Balderson DS, Papini RP (2004) A review of keratinocyte delivery to the wound bed. J Burn Care Rehabil 25:266–275

Choate KA, Kinsella TM, Williams ML et al (1996) Transglutaminase 1 delivery to lamellar ichthyosis keratinocytes. Hum Gene Ther 7:2247–2253

Cotsarelis G, Kaur P, Dhouilly D et al (1999) Epithelial stem cells in the skin: definition, markers, localization and functions. Exp Dermatol 8:80–88

Dellambra E, Vailly J, Pellegrini G et al (1998) Corrective transduction of human epidermal stem cells in laminin-5-dependent junctional epidermolysis bullosa. Hum Gene Ther 9:1359–1370

Dick JE (2003) Stem cells: self-renewal writ in blood. Nature 423:231–233

Dunnwald M, Tomanek-Chalkley A, Alexandrunas D et al (2001) Isolating a pure population of epidermal stem cells for use in tissue engineering. Exp Dermatol 10:45–54

Fortunel NO, Hatzfeld JA, Rosemary PA et al (2003) Long-term expansion of human functional epidermal precursor cells: promotion of extensive amplification by low TGF-beta1 concentrations. J Cell Sci 116:4043–4052

Freiberg RA, Choate KA, Deng H et al (1997) A model of corrective gene transfer in X-linked ichthyosis. Hum Mol Genet 6:927–933

Fuchs E,, Raghavan S (2002) Getting under the skin of epidermal morphogenesis. Nat Rev Genet 3:199–209

Fusenig NE, Amer SM, Boukamp P et al (1978) Characteristics of chemically transformed mouse epidermal cells in vitro and in vivo. Bull Cancer 65:271–279

Ghazizadeh S,, Taichman LB (2001) Multiple classes of stem cells in cutaneous epithelium: a lineage analysis of adult mouse skin. EMBO J 20:1215–1222

Goodell MA, Brose K, Paradis G et al (1996) Isolation and functional properties of murine hematopoietic stem cells that are replicating in vivo. J Exp Med 183:1797–1806

Gussoni E, Soneoka Y, Strickland CD et al (1999) Dystrophin expression in the mdx mouse restored by stem cell transplantation. Nature 401:390–394

Harrison DE (1980) Competitive repopulation: a new assay for long-term stem cell functional capacity. Blood 55:77–81

Heenen M,, Galand P (1997) The growth fraction of normal human epidermis. Dermatology 194:313–317

Hodgson GS,, Bradley TR (1979) Properties of haematopoietic stem cells surviving 5-fluorouracil treatment: evidence for a pre-CFU-S cell? Nature 281:381–382

Hulspas R, Quesenberry PJ (2000) Characterization of neurosphere cell phenotypes by flow cytometry. Cytometry 40:245–250

Ivanova NB, Dimos JT, Schaniel C et al (2002) A stem cell molecular signature. Science 298:601–604

Jahoda CA, Reynolds AJ, Oliver RF et al (1993) Induction of hair growth in ear wounds by cultured dermal papilla cells. J Invest Dermatol 101:584–590

Jones PH, Watt FM (1993) Separation of human epidermal stem cells from transit amplifying cells on the basis of differences in integrin function and expression. Cell 73:713–724

Kameda T, Nakata A, Mizutani T et al (2003) Analysis of the cellular heterogeneity in the basal layer of mouse ear epidermis: an approach from partial decomposition in vitro and retroviral cell marking in vivo. Exp Cell Res 283:167–183

Kaur P, Li A (2000) Adhesive properties of human basal epidermal cells: an analysis of keratinocyte stem cells, transit amplifying cells, and postmitotic differentiating cells. J Invest Dermatol 114:413–420

Kondo M, Weissman IL, Akashi K (1997) Identification of clonogenic common lymphoid progenitors in mouse bone marrow. Cell 91:661–672

Lapidot T, Sirard C, Vormoor J et al (1994) A cell initiating human acute myeloid leukaemia after transplantation into SCID mice. Nature 367:645–648

Lavker RM, Sun TT (1983) Epidermal stem cells. J Invest Dermatol 81 [1 Suppl]:121S–127S

Lavker RM, Sun TT (2000) Epidermal stem cells: properties, markers, and location. Proc Natl Acad Sci U S A 97:13473–13475

Lechner A, Leech CA, Abraham EJ et al (2002) Nestin-positive progenitor cells derived from adult human pancreatic islets of Langerhans contain side population (SP) cells defined by expression of the ABCG2 (BCRP1) ATP-binding cassette transporter. Biochem Biophys Res Commun 293:670–674

Lerner C, Harrison DE (1990) 5-Fluorouracil spares hemopoietic stem cells responsible for long-term repopulation. Exp Hematol 18:114–118

Li A, Pouliot N, Redevers R et al (2004) Extensive tissue-regenerative capacity of neonatal human keratinocyte stem cells and their progeny. J Clin Invest 113:390–400

Lyle S, Christofidou-Solomidou M, Liu Y et al (1998) The C8/144B monoclonal antibody recognizes cytokeratin 15 and defines the location of human hair follicle stem cells. J Cell Scie 111:3179–3188

Mackenzie IC (1997) Retroviral transduction of murine epidermal stem cells demonstrates clonal units of epidermal structure. J Invest Dermatol 109:377–383

Mackenzie IC, Bickenbach JR (1985) Label-retaining keratinocytes and Langerhans cells in mouse epithelia. Cell Tissue Res 242:551–556

Magnaldo T, Sarasin A (2002) Genetic reversion of inherited skin disorders. Mutat Res 509:211–220

Magnaldo T, Sarasin A (2004) Xeroderma pigmentosum: from symptoms and genetics to gene-based skin therapy. Cells Tissues Organs 177:189–198

Matic M, Evans WH, Brink PR et al (2002) Epidermal stem cells do not communicate through gap junctions. J Invest Dermatol 118:110–116

Matic M, Simon M (2003) Label-retaining cells (presumptive stem cells) of mice vibrissae do not express gap junction protein connexin 43. J Investig Dermatol Symp Proc 8:91–95

Michel M, Török N, Goodbout MJ et al (1996) Keratin 19 as a biochemical marker of skin stem cells in vivo and in vitro: keratin 19 expressing cells are differentially localized in function of anatomic sites, and their number varies with donor age and culture stage. J Cell Sci 109:1017–128

Mikkola HK, Klintman J, Yang H et al (2003) Haematopoietic stem cells retain long-term repopulating activity and multipotency in the absence of stem-cell leukaemia SCL/tal-1 gene. Nature 421:547–551

Morris RJ, Fischer SM, Sliga TJ et al (1985) Evidence that the centrally and peripherally located cells in the murine epidermal proliferative unit are two distinct cell populations. J Invest Dermatol 84:277–281

Morris RJ, Potten CS (1994) Slowly cycling (label-retaining) epidermal cells behave like clonogenic stem cells in vitro. Cell Prolif 27:279–289

Morris RJ, Tacker KC, Fischer SN et al (1988) Quantitation of primary in vitro clonogenic keratinocytes from normal adult murine epidermis, following initiation, and during promotion of epidermal tumors. Cancer Res 48:6285–6290

Morrison SJ, Weissman IL (1994) The long-term repopulating subset of hematopoietic stem cells is deterministic and isolatable by phenotype. Immunity 1:661–673

Niemann C, Watt FM (2002) Designer skin: lineage commitment in postnatal epidermis. Trends Cell Biol 12:185–192

Oshima H, Rochat A, Kedzia C et al (2001) Morphogenesis and renewal of hair follicles from adult multipotent stem cells. Cell 104:233–245

Park IK, Qian D, Kiel M et al (2003) Bmi-1 is required for maintenance of adult self-renewing haematopoietic stem cells. Nature 423:302–305

Pellegrini G, Dellambra E, Golisano O et al (2001) P63 identifies keratinocyte stem cells. Proc Natl Acad Sci U S America 98:3156–3161

Potten CS (1981) Cell replacement in epidermis (keratopoiesis) via discrete units of proliferation. Int Rev Cytol 69:271–318

Potten CS, Hendry JH (1973) Letter: clonogenic cells and stem cells in epidermis. Int J Radiat Biol Relat Stud Phys Chem Med 24:537–540

Potten CS, Morris RJ (1988) Epithelial stem cells in vivo. J Cell Sci Suppl 10:45–62

Ramalho-Santos M, Yoon S, Matsuzaki Y et al (2002) Stemness: transcriptional profiling of embryonic and adult stem cells. Science 298:597–600

Randall TD, Lund FE, Howard MC et al (1996) Expression of murine CD38 defines a population of long-term reconstituting hematopoietic stem cells. Blood 87:4057–4067

Reynolds AJ, Jahoda CA (1992) Cultured dermal papilla cells induce follicle formation and hair growth by transdifferentiation of an adult epidermis. Development 115:587–593

Ro S, Rannala B (2004) A stop-EGFP transgenic mouse to detect clonal cell lineages generated by mutation. EMBO Rep 5:914–920

Schmidt GH, Blount, Ponder BA et al MA (1987) Immunochemical demonstration of the clonal organization of chimaeric mouse epidermis. Development 100:535–541

Schneider TE, Barland C, Alex AM et al (2003) Measuring stem cell frequency in epidermis: a quantitative in vivo functional assay for long-term repopulating cells. Proc Natl Acad Sci U S A 100:11412–11417

Slack JM (2000) Stem cells in epithelial tissues. Science 287:1431–1433

Smith LG, Weissman IL, Heimfeld S et al (1991) Clonal analysis of hematopoietic stem-cell differentiation in vivo. Proc Natl Acad Sci U S A 88:2788–2792

Spangrude GJ, Heimfeld S, Weissman IL et al (1988) Purification and characterization of mouse hematopoietic stem cells. Science 241:58–62

Spangrude GJ, Johnson GR (1990) Resting and activated subsets of mouse multipotent hematopoietic stem cells. Proc Natl Acad Sci U S A 87:7433–7437

Sudo K, Ema H, Morita Y et al (2000) Age-associated characteristics of murine hematopoietic stem cells. J Exp Med 192:1273–1280

Szilvassy SJ, Humphries RK, Landsdorp PM et al (1990) Quantitative assay for totipotent reconstituting hematopoietic stem cells by a competitive repopulation strategy. Proc Natl Acad Sci U S A 87:8736–8740

Tani H, Morris RJ, Kaur P (2000) Enrichment for murine keratinocyte stem cells based on cell surface phenotype. Proc Natl Acad Sci U S A 97:10960–10965

Taswell C (1981) Limiting dilution assays for the determination of immunocompetent cell frequencies. I Data analysis. J Immunol 126:1614–1619

Taylor G, Lehrer MS, Jensen PJ et al (2000) Involvement of follicular stem cells in forming not only the follicle but also the epidermis. Cell 102:451–461

Terskikh AV, Miyamoto T, Chang C et al (2003) Gene expression analysis of purified hematopoietic stem cells and committed progenitors. Blood 102:94–101

Terunuma A, Jackson KL, Kapoor V et al (2003) Side population keratinocytes resembling bone marrow side population stem cells are distinct from label-retaining keratinocyte stem cells. J Invest Dermatol 121:1095–1103

Trempus CS, Morris RJ, Bortner CD et al (2003) Enrichment for living murine keratinocytes from the hair follicle bulge with the cell surface marker CD34. J Invest Dermatol 120:501–511

Trevisan M, Iscove NN (1995) Phenotypic analysis of murine long-term hemopoietic reconstituting cells quantitated competitively in vivo and comparison with more advanced colony-forming progeny. J Exp Med 181:93–103

Triel C, Vestergaard ME, Bolund L et al (2004) Side population cells in human and mouse epidermis lack stem cell characteristics. Exp Cell Res 295:79–90

Tumbar T, Guasch G, Greco V et al (2004) Defining the epithelial stem cell niche in skin. Science 303:359–363

Van Zant G (1984) Studies of hematopoietic stem cells spared by 5-fluorouracil. J Exp Med 159:679–690

Wan H, Stone MG, Simpson C et al (2003) Desmosomal proteins, including desmoglein 3, serve as novel negative markers for epidermal stem cell-containing population of keratinocytes. J Cell Sci 116:4239–4248

Wang CK, Nelson CF, Brinnkman AM et al (2000) Spontaneous cell sorting of fibroblasts and keratinocytes creates an organotypic human skin equivalent. J Invest Dermatol 114:674–680

Weissman IL (2002) The road ended up at stem cells. Immunol Rev 185:159–174

Wilke MS, Edens M, Scott RE et al (1988) Ability of normal human keratinocytes that grow in culture in serum-free medium to be derived from suprabasal cells. J Natl Cancer Inst 80:1299–304

Zhang W, Remenyik E, Zelterman D et al (2001) Escaping the stem cell compartment: sustained UVB exposure allows p53-mutant keratinocytes to colonize adjacent epidermal proliferating units without incurring additional mutations. Proc Natl Acad Sci U S A 98:13948–13953

Zhou JX, Jia LW, Yang YJ et al (2004) Enrichment and characterization of mouse putative epidermal stem cells. Cell Biol Int 28:523–529

5 Corneal Cells for Regeneration

S. Kinoshita, T. Nakamura

Abstract. In cases of corneal epithelial stem cell deficiency where ocular surface reconstruction is required, corneal epithelial replacement using a tissue engineering technique shows great potential. Autologous cultivated corneal epithelial stem cell sheets are the safest and most reliable forms of sheet we can use for such treatment; however, they are not useful for treating bilaterally affected ocular surface disorders. In order to treat such cases, we must choose either an allogeneic cultivated corneal epithelial sheet or an autologous cultivated oral mucosal epithelial sheet. If we use the former, the threat of immunological reaction must be dealt with. Therefore, it is imperative that we have a basic understanding of the immunological aspects of ocular surface reconstruction using allogeneic tissues. When using an autologous cultivated oral mucosal epithelial sheet, a basic understanding of ocular surface epithelial biology is required as the sheet is not exactly the same as corneal epithelium.

5.1 Introduction

The concept of an "ocular surface" is widely recognized in the field of ophthalmology and our understanding of the role of ocular surface biology and immunology has greatly improved due to the numerous research studies carried out in this field (Thoft and Friend 1979). Over the past 20 years, there have been several scientific discoveries such as the identification of corneal epithelial stem cells, the establishment of novel methods in epithelial culturing and the understanding of extracellular matrices which have enabled us to adopt a novel surgical approach to treat ocular surface disorders using regenerative medicine. Regenerative medicine through tissue engineering is a newly developed area of medicine which uses somatic stem cells to generate biological substitutes and improve tissue functions (Langer and Vacanti 1993). Success depends on two important factors: stem cells and the extracellular matrices. In cases of severely affected ocular surface disorders, we are now able to produce cultivated corneal or cornea-like epithelial sheets in vitro and transplant them onto the ocular surface. Here, we summarize this concept and the clinical application of cultivated mucosal epithelial transplantation for ocular surface disorders.

5.2 The Concept of Cellular Surgery

5.2.1 Ocular Surface Disorders

Ocular surface integrity is maintained by corneal and conjunctival epithelia. These make up the normal ocular surface and each has a distinct cellular phenotype. These epithelia, accompanied by an intact tear film, ensure the ocular surface integrity. The corneal epithelium, in particular, plays a vital part in maintaining the cornea's transparency and avascularity. On the basis of numerous investigations, it is believed that the corneal epithelial stem cells exist in the basal cell layer of the limbal region where, in normal human subjects, palisades of Vogt are seen (Schermer et al., 1986; Kinoshita et al., 2001). Persistent epithelial defects, such as conjunctivalization with vascularization, keratinization, scarring, etc., with associated profound visual loss are caused by severe damage to the limbal region. In other words, corneal epithelial stem cell deficiency seriously affects the corneal surface (Fig. 1).

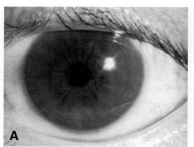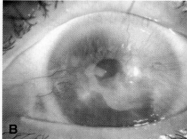

Fig. 1a, b. The normal ocular surface (**a**) and severe ocular surface disorders (chemical injured eye, **b**)

5.2.2 Background and Biological Aspect

Over the past 20 years, novel surgical modalities, designed to reconstruct diseased or damaged ocular surface epithelia, have been developed. The concept of ocular surface reconstruction was first introduced via an autologous conjunctival transplantation for unilateral chemical injury reported (Thoft 1977). Thoft described conjunctival transplantation for unilaterally affected chemical injuries; the surgery was performed by removing scar tissue from a patient's corneal surface and placing four pieces of conjunctival autograft, taken from the contralateral eye at the limbus, in order to resurface the cornea by regenerating conjunctival epithelial cells from these autografts. Thoft also described a similar surgical technique, keratoepithelioplasty. However, this technique employed a different tissue source, donor corneal lenticlules, to regenerate corneal epithelial cells. Although the concept of corneal epithelial stem cells was not established at that time, the biological differences between regenerated corneal and conjunctival epithelia were known. Over time, keratoepithelioplasty, as a form of treatment, has gradually been accepted, despite initial disputes amongst researchers caused by the fact that it is a form of epithelial allograft (Kaufman 1984). In fact, keratoepithelioplasty has proved to be dramatically effective in treating peripheral corneal ulcers, including Mooren's ulcer (Kinoshita et al. 1991), since it not only supplies a regenerated corneal epithelium but also an appropriate corneal substrate for inhibiting conjunctival invasion onto the cornea. Sun's group proposed the stem cell concept (Schermer et al. 1986), which had a tremendous impact on keratoepithelioplasty, leading to limbal autograft transplantation (Kenyon and Tseng 1989). Tsai and Tseng then introduced limbal allograft transplantation aimed at achieving the permanent life-span of regenerated corneal epithelium via stem cell transplantation, although intensive immunosuppressive therapy was necessary (Tsai and Tseng 1994). These surgical procedures are classed as 'cellular surgery', a form of primitive regenerative medicine, as they are a kind of in vivo expansion of corneal epithelial cells. Significntly, in 1995, Kim and Tseng reported that amniotic membrane (AM) transplantation was capable of inhibiting subepithelial scarring in ocular surface reconstruction (Kim and Tseng 1995). Since that time, AM transplantation combined with limbal allografts has been used to treat certain challenging severe ocular surface disorders such as Stevens-Johnson syndrome and ocular cicatricial pemphigoid (Tsubota et al. 1996). A well-summarized review of these procedures and clinical results has been reported (Holland and Schwartz 1996). As a further advancement, the transplantation of a corneal or oral epithelial sheet, free of subepithelial tissue, was challenged via animal experimentation (Gipson et al., 1985, 1986).

5.3   Transplantation of Cultivated Mucosal Epithelial Cell Sheets

While corneal epithelial transplantations, including limbal autografts and limbal allografts, have undoubtedly helped to improve the outcome of ocular surface reconstruction in a number of situations, a limbal autograft requires a fairly large amount of limbal tissue from the healthy eye and cannot be considered if the disease or injury is bilateral. Thus, in order to improve the surgical results of ocular surface reconstruction for severe stem cell deficiencies, cultivated corneal epithelial sheets need to be developed in vitro from a small portion of limbal epithelium.

5.3.1   The History of Creating Mucosal Epithelial Sheets

In the initial stages of ocular surface epithelial surgery, in the 1980s, there were at least two avenues of investigation that sought to create a mucosal epithelial sheet. One was the direct sampling of either corneal or oral mucosal epithelial sheets using dispase and gentle mechanical treatment. The corneal epithelial sheet, which was taken directly from in vivo cornea attached itself to the corneal stroma in animal experiments, but it peeled away easily during lid movement (Gipson and Grill 1982). On the other hand, the oral mucosal epithelial sheet remained on the cornea, but induced severe neovascularization (Gipson et al. 1986).

The other method was to establish a cultivated epithelial cell sheet. In the 1970s, Green (Rheinwald and Green 1975) was successful in establishing a procedure to constitute the biological epithelial sheet of cultured human keratinocytes from the epidermis. This method has since been used in dermatological fields. Thus, attention has been focused on the ex vivo expansion of corneal epithelial cells, i.e. a cultivated corneal epithelial sheet. Firstly, the cultivation of corneal epithelial cells on the basement membrane of scraped rabbit corneal stroma was examined (Friend et al. 1982). Several other substrates such as hydrogel coated with fibronectin, and collagen matrix were also investigated in order to develop this procedure (Kobayashi and Ikada 1991; Minami et al. 1993). Yet, it was not until 1997 that clinical reports regarding successful ocular surface reconstruction using the ex vivo expansion of limbal epithelial cells were recorded. This was probably due to the lack of understanding surrounding corneal epithelial stem cell cultivation and/or its proper substrate. Some groups tried to reconstruct not only the epithelial sheet, but also three layers of corneal tissue – a 'corneal equivalent' – using cell-line cells supported by natural and synthetic polymers (Griffith et al. 1999). This kind of corneal equivalent is now ready to be used for testing toxicity and drug efficacy but not for clinical use because of immortalized cell lines used and other factors related to carrier materials.

5.3.2   Cultivated Corneal Epithelial Stem Cell Sheet Transplantation

The first successful ocular surface reconstruction using autologous cultivated corneal epithelial stem cells in patients with severely affected unilateral ocular surface disease was reported (Pellegrini et al. 1997). Pelligrini's group developed a method to reconstruct stratified corneal epithelial cell sheets on petrolatum gauze or a soft contact lens as a carrier and treated two patients who had unilateral chemical burns by transplanting cultivated corneal epithelial cells taken from the limbus of the healthy contralateral eye. It seems that this was successful because they adopted Rheinwald and Green's well-recognized keratinocyte-culturing method, incorporating the use of 3T3 feeder layers to sustain epithelial stem cells.

Following the establishment of a good substrate, several researchers investigated the option of using amniotic membrane as a carrier (Tsai et al. 2000; Schwab et al. 2000). As a result of their findings, researchers soon realized amniotic membrane's (AM) potential to act as a carrier for corneal epithelial cell cultures.

Amniotic Membrane as an Epithelial Carrier

In the past, AM was used, with or without limbal transplantation, in a range of ocular surgeries and it proved to be useful for the treatment for thermal and chemical injuries, severe pterygium, persistent or deep corneal ulcers, ocular cicatricial pemphigoid, Stevens-Johnson syndrome, and other limbal stem cell deficiencies (Dua and Azuara-Blanco 1999).

AM is the innermost layer of the fetal membrane. It is composed of a monolayer of amniotic epithelial cells, a thick basement membrane, and an avascular stroma (Fig. 2). Various mechanisms of AM in transplantation have been reported:

 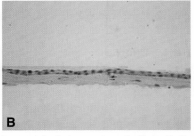

Fig. 2a, b. The human amniotic membrane (**a**). It is composed of amniotic epithelial cells, a thick basement membrane and an avascular stroma (**b**)

rescein just after being transferred onto the ocular surface. Furthermore, at 48 h after transplantation, there was no epithelial damage to the transplanted corneal epithelium. Visual acuity was not affected by the transplanted amniotic membrane and clarity increased day by day. Surprisingly, in all of the acute-phase patients, the preoperative ocular surface inflammation which had not been controlled by conventional treatment including corticosteroid, cyclosporine, therapeutic soft contact lens, as well as the amniotic membrane patch, rapidly decreased after surgery.

In the three different diseases mentioned above, the long-term visual prognosis and epithelial stability in the chronic-phase patients were varied. In the case of severe chemical injuries, the transplanted corneal epithelium was clear and stable even to 4 years after transplantation, and there was only a small amount or indeed no conjunctival inflammation during the whole of the postoperative period. On the other hand, in Stevens-Johnson syndrome patients, mild to moderate ocular surface inflammation occurred several months after cultivated corneal epithelial transplantation, and then decreased within 18 months postoperatively. Whereas subconjunctival fibrosis had not progressed in the Stevens-Johnson syndrome, conjunctival scarring such as symblepharon and shortening of the fornix had progressed in ocular cicatricial pemphigoid. In most of the chronic-phase patients with immunological abnormalities such as Stevens-Johnson syndrome and ocular cicatricial pemphigoid, the phenotypes of ocular surface cells on AM gradually changed from donor to host epithelial cells over a couple of years; however, subepithelial scarring and neovascularization did not progress. This phenomenon is considered, in part, to be due to a mild rejection of the transplanted corneal epithelial cells. Although graft survival was not very long in some eyes of these chronic cases, the ocular surface maintained its transparency and the patients obtained a better visual function compared to their condition before undergoing cultivated corneal epithelial transplantation.

Interestingly, we noted that at the time of second surgery the rejected transplants were easily removed from patients' eyes and the exposed stroma were fairly transparent with less scarring changes (Nakamura et al. 2003a). Also, AM was detected in the removed samples taken at second surgery, which may play a part in inhibiting scarring changes. Although methicillin-resistant *Staphylococcus aureus* (MRSA) or methicillin-resistant *Staphylococcus epidermidis* (MRSE) infection occurred with high frequency in the Stevens-Johnson syndrome patients, most cases healed with little corneal scarring (Sotozono et al. 2002). We found that persistent epithelial defects occurred in some eyes with chronic ocular cicatricial pemphigoid, but corneal perforation did not occur and we were successful in achieving epithelialization via conjunctivalization.

The use of autologous tissue-engineered corneal epithelial replacements (with the harvesting of a minimal biopsy in order to prevent damage to the uninjured donor eye), is the ideal therapy for unilateral ocular surface diseases (OSD) such as chemical and thermal injuries. With this in mind, several researchers have reported that autologous cultivated corneal limbal epithelial cells from uninvolved eyes could be used for effective ocular surface reconstruction in patients with unilateral OSD (Tsai et al. 2000; Schwab et al. 2000; Nakamura et al. 2004a). These reports offer hope to patients with unilateral lesions or severe ocular surface damage. Meanwhile, we posit that the cultivation of AM on corneal limbal epithelial cells harvested from minimal specimens is possible, but rather difficult to achieve using conventional culturing techniques; we have to further develop the culturing system to produce a sufficiently stratified, well-differentiated epithelium that will cover the entire corneal surface. In addition, a longer follow-up period is required in order to ascertain the stability of the ocular surface in autologous cultivated corneal epithelial transplantation.

5.3.3   Transplantation of a Cultivated Oral Mucosal Epithelial Sheet

Background

As severe ocular surface disorders are mostly bilateral, we have to perform allogeneic corneal epithelial transplantation, but in order to prevent allograft rejection, we have to use prolonged postoperative immunosuppressive therapy. This markedly reduces the patients' quality of life and severely affects clinical results. Thus we have investigated whether the ocular surface could be reconstructed using an autologous mucosal epithelium of nonocular surface origin. Hence, we have attempted to

overcome the problems involved with allogeneic transplantation by using cultivated oral epithelial cells as a substitute for corneal epithelial
cells (Nakamura et al. 2003b, 2003c).

In the past, several researchers investigated the possibility of using
oral mucosa for ocular surface reconstruction. Ballen reported that oral
mucosal grafts, which included both epithelium and subepithelial tissues, heavily vascularized with early fibrosis (Ballen 1963). Also Gipson
et al. reported that the in vivo oral epithelium, freed of underlying connective tissue, was not maintained in central avascular corneal regions
(Gipson et al. 1986).

Animal Experiments

Considering these findings, we have cultured rabbit oral mucosal epithelial cells on AM as a carrier, using a modified version of our previously
established culture method. Small oral biopsies (approximately 2–3 mm)
were obtained from the oral cavity under local anesthesia and were then
incubated with enzymatic reagents such as dispase and trypsin-EDTA to
separate the cells from the underlying connective tissue. The resultant
single-cell suspensions of oral mucosal epithelial cells were co-cultured
for 2–3 weeks on a denuded AM carrier, with inactivated 3T3 fibroblast.
An air-lifting technique was used towards the end of the culture period in order to facilitate epithelial differentiation and stratification. The
oral mucosal epithelial cells cultivated on AM showed five to six layers
of stratification and appeared very similar to in vivo normal corneal
epithelium (Fig. 5). Immunohistochemistry, used for several keratins,
showed the presence of nonkeratinized, stratified-specific keratins 4/13
and cornea-specific 3 in the cultivated oral epithelial sheets. In contrast, epidermal keratinization-related keratin 1/10 and cornea-specific
keratin 12 were not expressed in any layers of the oral epithelial cells.
Electron microscopic examination showed that the cultivated oral epithelial sheet on the AM had junctional structures, such as desmosomal,
hemi-desmosomal and tight junctions that were almost identical to those
of in vivo corneal epithelial cells. Based on this information, we have
concluded that oral epithelial cells cultivated on denuded AM have the
potential ability to differentiate into cornea-like epithelia under culture
conditions (Nakamura et al. 2003b, 2003c).

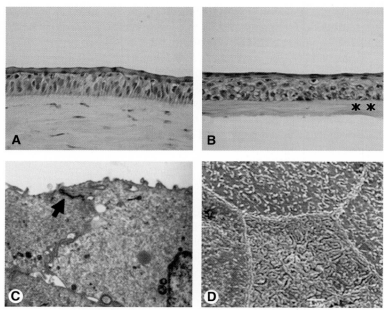

Fig. 5a–b. Light micrographs showing cross-sections of the rabbit normal corneal epithelial cells (**a**) and cultivated oral epithelial cells on amniotic membrane (**b**) stained with hematoxylin and eosin. The cultivated oral epithelial sheet had five to six layers of stratified, well-differentiated cells and appeared very similar to normal corneal epithelium. Transmission and scanning electron micrographs of cultivated oral epithelial cells on denuded amniotic membrane showed that the cells appear healthy and well-formed with distinct cell boundaries (**d**). What appear to be tight junctions (*arrows*) are occasionally evident between the most superficial cell layers (**c**). *Asterisks*, amniotic stroma

Following the successful culturing of oral mucosal epithelial cells on AM, we tried to reconstruct damaged ocular surfaces using a rabbit model mimicking a severe ocular surface disease, using superficial keratectomy. The ocular surface was then reconstructed with an autologous cultivated oral mucosal epithelial sheet. At 48 h after surgery, most of the area of the cultivated oral mucosal epithelial sheet transplanted possessed an intact epithelium. At 10 days after transplantation, the ocular surface, covered by the transplanted epithelium, was intact and without defects, suggesting that the autologous transplantation of cultivated oral mucosal epithelia is a viable procedure for ocular surface reconstruction.

Clinical Experience

On the basis of these important findings, we successfully applied this method to patients with severe ocular surface disorders and reconstructed the ocular surfaces (Nakamura et al. 2004b) (Fig. 6). Using this unique tissue engineering procedure generated from autologous oral mucosa,

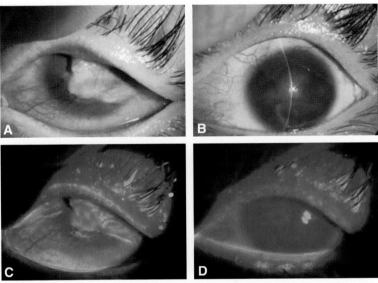

Fig. 6a–d. a–d Representative slit-lamp photographs taken before transplantation without (**a**) and with fluorescein (**b**). The photographs in **c** and **d** were taken at the 11 months after transplantation without (**c**) and with fluorescein (**d**). Before transplantation, the eye manifested inflammatory subconjunctival fibrosis with neovascularization, conjunctivalization, and severe symblepharon. Eleven months after transplantation, the corneal surface was stable without defects. (Reproduced with permission of the British Journal of Ophthalmology from Nakamura et al. 2004)

we can avoid administering the intensive immunosuppressive medication, thus reducing the risk of postoperative complications such as infections, steroid induced glaucoma and cataracts. In our initial clinical trials, we used these cultivated oral mucosal epithelial sheets to perform two different forms of surgery: to reconstruct the corneal surface of a severe bilateral corneal stem cell deficiency instead of using allogeneic corneal epithelium and to reconstruct the conjunctival fornix in patients with severe symblepharon formation associated with ocular cicatricial pemphigoid, Stevens-Johnson syndrome and chemical or thermal burns.

In order to prepare the cultivated oral mucosal epithelial sheet, $2–3$ mm^2 oral mucosal biopsy specimens were obtained from our patients 3 weeks prior to transplantation. Using the same method carried out on the animal model, single-cell suspensions of oral mucosal epithelial cells were cultured on acellular AM. As in the case of the animal model, the cultivated oral mucosal epithelium consisted of five to six nonkeratinized epithelial layers with normal transparency just prior to surgery. The surgical procedure is practically the same as that for cultivated corneal epithelial sheet transplantation. Using fluorescein staining, we observed an epithelial phenotype of transplanted cultivated oral mucosal epithelium which was somewhat distinguishable from the conjunctival epithelium. Our preliminary data show the successful survival of autologous cultivated oral mucosal epithelium on the ocular surface without returning to an in vivo oral tissue phenotype, as was the case with oral mucosal transplantation in the past. We can explain this major difference as being due to the elimination of subepithelial fibrous tissue and vascular components in oral mucosa during the tissue culturing system. It is possible that amniotic membrane has some effect on this phenomenon as well. One adverse effect of this procedure was that the transplanted cultivated oral mucosal epithelium sometimes showed a neovascularization, to some extent, in the peripheral cornea with epithelial thickening. Clinical investigations are now under way in order to understand this phenomenon.

Cultivated oral mucosal sheets on AM were also used to reconstruct the conjunctival fornix; this technique was successful in treating ocular surface diseases, including ocular cicatricial pemphigoid, chemical injury, etc. However, abnormal postoperative fibrovascular proliferation caused by primary diseases must be considered, as it is still critical to the long-term prognosis. So, this tissue-engineering process involves the application of nonocular mucosal epithelium in order to reconstruct an ocular surface, providing an acceptable level of comfort and visual rehabilitation for visually impaired patients. In addition, Nishida and colleges reported the successful corneal reconstruction with carrier-free cultivated oral epithelial sheet in patients with chronic phase of severe ocular surface disease (Nishida et al. 2004).

5.4   Future Goal

In conclusion, the cultivated corneal epithelial transplantation technique holds many future possibilities. For example, when a selected donor–recipient combination is required, such as in MHC matching, the cultivated corneal epithelial transplantation method will be helpful, as the cell culture period affords adequate time to select a recipient for the donor. Furthermore, there would be more possibility for master stem cell banking. Also, a more sophisticated way of culturing oral mucosa or conjunctiva may provide us with a better quality epithelial sheet for autologous transplantation.

Considering all of these possibilities, it is our belief and hope that more advanced strategies will be developed in the near future to further enhance the treatment of ocular surface disorders.

Acknowledgements. Supported by grants from the Japanese Ministry of Health and Welfare and the Japanese Ministry of Education, Tokyo, the Kyoto Foundation for the Promotion of Medical Science, and the Intramural Research Fund of the Kyoto Prefectural University of Medicine.

References

Ballen PH (1963) Mucous membrane grafts in chemical (lye) burns. Am J Ophthalmol 55:302–312

Ban Y, Cooper LJ, Fullwood NJ, Nakamura T, Tsuzuki M, Koizumi N, Dota A, Mochida C, Kinoshita S (2003) Comparison of ultrastructure, tight junction-related protein expression and barrier function of human corneal epithelial cells cultivated on amniotic membrane with and without air-lifting. Exp Eye Res 76:735–743

Cotsarelis G, Cheng SZ, Dong G, Sun TT, Lavker RM (1989) Existence of slow-cycling limbal epithelial basal cells that can be preferentially stimulated to proliferate: implications on epithelial stem cells. Cell 57:201–209

Dua HS, Azuara-Blanco A (1999) Amniotic membrane transplantation. Br J Ophthalmol 83:748–752

Endo K, Nakamura T, Kawasaki S, Kinoshita S (2004) Human amniotic membrane, like corneal epithelial basement membrane, manifests the a5 chain of type IV collagen. Invest Ophthalmol Vis Sci. 45:1771–1774

Friend J, Kinoshita S, Thoft RA, Eliason JA (1982) Corneal epithelial cell cultures on stromal carriers. Invest Ophthalmol Vis Sci 23:41–49

Fukuda K, Chikama T, Nakamura M, Nishida T (1999) Differential distribution of subchains of the basement membrane components type IV collagen and laminin among the amniotic membrane, cornea and conjunctiva. Cornea 18:73–79

Gipson IK, Grill SM (1982) A technique for obtaining sheets of intact rabbit corneal epithelium. Invest Ophthalmol Vis Sci 23:269–273

Gipson IK, Friend J, Spurr SJ (1985) Transplantation of corneal epithelium to rabbit corneal wounds in vivo. Invest Ophthalmol Vis Sci 26:425–233

Gipson IK, Geggel HS, Spurr-Michaud SJ (1986) Transplant of oral mucosal epithelium to rabbit ocular surface wounds in vivo. Arch. Ophthalmol 104:1529–1533

Griffith M, Osborne R, Munger R, Xiong X, Doillon CJ, Laycock NL, Hakim M, Song Y, Watsky MA (1999) Functional human corneal equivalents constructed from cell lines. Science 286:2169–2172

Grueterich M, Espana E, Tseng SC (2002) Connexin 43 expression and proliferation of human limbal epithelium on intact and denuded amniotic membrane. Invest Ophthalmol Vis Sci 43:63–71

Holland EJ, Schwartz GS (1996) The evolution of epithelial transplantation for severe ocular surface disease and a proposed classification system. Cornea 15:549–556

Kaufman HE (1984) Keratoepithelioplasty for the replacement of damaged corneal epithelium. Am J Ophthalmol 97:100–101

Kenyon KR, Tseng SC (1989) Limbal autograft transplantation for ocular surface disorders. Ophthalmology 96:709–722

Kim JC, Tseng SC (1995) Transplantation of preserved human amniotic membrane for surface reconstruction in severely damaged rabbit corneas. Cornea 14:473–484

Kim JS, Kim JC, Na BK, Jeong JM, Song CY (2000) Amniotic membrane patching promotes healing and inhibits protease activity on wound healing following acute corneal alkali burn. Exp Eye Res 70:329–337

Kinoshita S, Ohashi Y, Ohji M, Manabe R (1991) Long-term results of kera-toepithelioplasty in Mooren's ulcer. Ophthalmology 98:438–445

Kinoshita S, Adachi W, Sotozono C, Nishida K, Yokoi N, Quantock AJ, Okubo K (2001) Characteristics of the human ocular surface epithelium. Prog Retin Eye Res 20:639–673

Kobayashi H, Ikada Y (1991) Corneal cell adhesion and proliferation on hydro-gel sheets bound with cell-adhesive proteins. Curr Eye Res 10:899–908

Koizumi N, Inatomi T, Sotozono C, Fullwood NJ, Quantock AJ, Kinoshita S (2000a) Growth factor mRNA and protein in preserved human amniotic membrane. Curr Eye Res 20:173–177

Koizumi N, Inatomi T, Quantock AJ, Fullwood NJ, Dota A, Kinoshita S (2000b) Amniotic membrane as a substrate for cultivating limbal corneal epithelial cells for autologous transplantation in rabbits. Cornea 19:65–71

Koizumi N, Fullwood NJ, Bairaktaris G, Inatomi T, Kinoshita S, Quantock AJ (2000c) Cultivation of corneal epithelial cells on intact and denuded human amniotic membrane. Invest Ophthalmol Vis Sci 41:2506–2513

Koizumi N, Inatomi T, Suzuki T, Sotozono C, Kinoshita S (2001a) Cultivated corneal epithelial transplantation for ocular surface reconstruction in acute phase of Stevens-Johnson syndrome. Arch Ophthalmol 119:298–300

Koizumi N, Inatomi T, Suzuki T, Sotozono C, Kinoshita S (2001b) Cultivated corneal epithelial stem cell transplantation in ocular surface disorders. Oph-thalmology 108:1569–1574

Koizumi N, Cooper LJ, Fullwood NJ, Nakamura T, Inoki K, Tsuzuki M, Ki-noshita S (2002) An evaluation of cultivated corneal limbal epithelial cells, using cell-suspension culture. Invest Ophthalmol Vis Sci 43:2114–2121

Langer R, Vacanti JP (1993) Tissue engineering. Science 14:920–926

Lee, SB Li DQ, Tan DT, Meller DC, Tseng SCG (2000) Suppression of TGF-beta signalling in both normal conjunctival fibroblasts and pterygial body fibroblasts by amniotic membrane. Curr Eye Res 20:325–334

Meller D, Pires RT, Tseng SC (2002) Ex vivo preservation and expansion of human limbal epithelial stem cells on amniotic membrane cultures. Br J Ophthalmol 86:463–471

Minami Y, Sugihara H, Oono S (1993) Reconstruction of cornea in three-dimensional collagen gel matrix culture. Invest Ophthalmol Vis Sci 34:2316–2324

Nakamura T, Koizumi N, Tsuzuki M, Inoki K, Sano Y, Sotozono C, Kinoshita, S (2003a) Successful regrafting of cultivated corneal epithelium using amni-otic membrane as a carrier in severe ocular surface disease. Cornea 22:70–71

Nakamura T, Endo K, Cooper L, Fullwood NJ, Tanifuji N, Tsuzuki M, Koizumi N, Inatomi T, Sano Y, Kinoshita S (2003b) The successful culture and autologous transplantation of rabbit oral mucosal epithelial cells on amniotic membrane. Invest Ophthalmol Vis Sci 44:6–16

Nakamura T, Kinoshita S (2003c) Ocular surface reconstruction using cultivated mucosal epithelial stem cells. Cornea 22:S75–S80

Nakamura T, Inatomi T, Sotozono C, Koizumi N, Kinoshita S (2004a) Successful primary culture and autologous transplantation of corneal limbal epithelial cells from minimal biopsy for unilateral severe ocular surface disease. Acta Ophthalmol Scan 82:468–471

Nakamura T, Inatomi T, Sotozono C, Amemiya T, Kanamura N, Kinoshita S (2004b) Transplantation of cultivated autologous oral mucosal epithelial cells in patients with severe ocular surface disorders. Br J Ophthalmol 88:1280–1284

Nishida K, Yamato M, Hayashida Y, Watanabe K, Maeda N, Watanabe H, Yamamoto K, Nagai S, Kikuchi A, Tano Y, Okano T (2003) Functional bioengineered corneal epithelial sheet grafts from corneal stem cells expanded ex vivo on a temperature-responsive cell culture surface. Transplantation 77:379–385

Nishida K, Yamato M, Hayashida Y, Watanabe K, Yamamoto K, Adachi E, Nagai S, Kikuchi A, Maeda N, Watanabe H, Okano T, Tano Y (2004) Corneal reconstruction with tissue-engineered cell sheets composed of autologous oral mucosal epithelium. N Engl J Med 351:1187–1196

Park WC, Tseng SCG (2000) Modulation of acute inflammation and keratocyte death by suturing, blood, and amniotic membrane in PRK. Invest Ophthalmol Vis Sci 41:2906–2914

Pellegrini G, Traverso CE, Franzi AT, Zingirian M, Cancedda R, De Luca M (1997) Long-term restoration of damaged corneal surfaces with autologous cultivated corneal epithelium. Lancet 349:990–993

Rama P, Bonini S, Lambiase A, Golosano O, Paterna P, De Luca M, Pellegrini, G (2001) Autologous fibrin-cultured limbal stem cells permanently restore the corneal surface of patients with total limbal stem cell deficiency. Transplantation 72:1478–1485

Rheinwald JG, Green H (1975) Serial cultivation of strains of human epidermal keratinocytes: the formation of keratinizing colonies from single cells. Cell 6:331–344

Schermer A, Galvin S, Sun T (1986) Differentiation-related expression of a major 64 K corneal keratin in vivo and in culture suggests limbal location of corneal epithelial stem cells. J Cell Biol 103:49–62

Schwab IR, Reyes M, Isseroff R (2000) Successful transplantation of bioengineered tissue replacements in patients with ocular surface disease. Cornea 19:421–426

Solomon A, Rosenblatt M, Monroy D, Ji Z, Pflugfelder SC, Tseng SCG (2001) Suppression of interleukin 1α and interleukin 1β in human limbal epithelial cells cultured on the amniotic membrane stromal matrix. Br J Ophthalmol 85:444–449

Sotozono C, Inagaki K, Fujita A, Koizumi N, Sano Y, Inatomi T, Kinoshita, S (2002) Methicillin-resistant Staphylococcus aureus and methicillin-resistant Staphylococcus epidermidis infections in the cornea. Cornea 21:S94–S101

Thoft RA (1977) Conjunctival transplantation. Arch Ophthalmol 95:1425–1427

Thoft RA, Friend J (eds) (1979) The ocular surface. Int Ophthalmol Clin 19: 283 pp.

Thoft RA (1984) Keratoepithelioplasty. Am J Ophthalmol 97:1–6

Tsai RJF, Tseng SCG (1994) Human allograft limbal transplantation for corneal surface reconstruction. Cornea 13:389–400

Tsai RJF, Li LM, Chen JK (2000) Reconstruction of damaged corneas by transplantation of autologous limbal epithelial cells. N Engl J Med 343:86–93

Tseng SCG, Li DQ, Ma X (1999) Suppression of transforming growth factor-beta isoforms, TGF-beta receptor type II, and myofibroblast differentiation in cultured human corneal and limbal fibroblasts by amniotic membrane matrix. J Cell Physiol 179:325–335

Tsubota K, Satake Y, Ohyama M, Toda I, Takano Y, Ino M, Shinozaki N, Shimazaki J (1996) Surgical reconstruction of the ocular surface in advanced ocular cicatricial pemphigoid and Stevens-Johnson syndrome. Am J Ophthalmol 122:38–52

6 Polycomb Gene Product Bmi-1 Regulates Stem Cell Self-Renewal

H. Nakauchi, H. Oguro, M. Negishi, A. Iwama

Abstract. The Polycomb group (PcG) gene $Bmi\text{-}1$ has recently been implicated in the maintenance of hematopoietic stem cells (HSCs). However, the role of each component of PcG complex in HSCs and the impact of forced expression of PcG genes on stem cell self-renewal remain to be elucidated. To address these issues, we performed both loss-of-function and gain-of-function analysis on various PcG proteins. Expression analysis revealed that not only $Bmi\text{-}1$ but also other PcG genes are predominantly expressed in HSCs. Loss-of-function analyses, however, demonstrated that absence of $Bmi\text{-}1$ is preferentially linked with a profound defect in HSC self-renewal, indicating a central role for Bmi-1, but not the other components, in the maintenance of HSC self-renewal. Over-expression analysis of PcG genes also confirmed an important role of Bmi-1 in HSC self-renewal. Our findings indicate that the expression level of Bmi-1 is the critical determinant for the self-renewal capacity of HSCs. These findings uncover novel aspects of stem cell regulation exerted through epigenetic modifications by the PcG proteins.

6.1   Introduction

Stem cells have been expected to be limitless sources for tissue or organ regeneration because of their self-renewal and mutilineage differentiation capacities. Hematopoietic stem cells (HSCs) supply all blood cells throughout life utilizing their self-renewal and multilineage differentiation capabilities. However, self-renewal capacity, despite being the hallmark of stemness, has never been understood in molecular terms for any type of adult stem cells, including HSCs. There must be stem cell-specific gene expression patterns stabilized by changes in chromatin structure that are maintained through cell divisions by the counteractions of transcriptional activatiors of the trithorax groupt (TrxG) proteins and repressors of the polycomb group (PcG) (Jacobs and van Lohuizen 2002; Orland 2003). PcG proteins form multiprotein complexes that play an important role in the maintenance of transcriptional repression of target genes. At least two distinct PcG complexes have been identified and well characterized. One complex includes Eed, EzH1, and EzH2, and the other includes Bmi-1, Mel-18, Mph1/Rae28, M33, Scmh1, and Ring1A/B. Eed-containing complexes control gene repression through recruitment of histone deacetylase followed by local chromatin deacetylation, and by methylation of histone H3 Lysine 27 by EzH2. Bmi-1-complexes are recruited to methylated histone H3 Lysine 27 via M33 chromodomain to contribute to the static maintenance of epigenetic memory (Fischle et al. 2003). In addition, Ring1B in Bmi-1-containing complex has recently been identified as a ubiquitin ligase for histone H2A (Wang et al. 2004). These two types of complexes coordinately maintain positional memory along the anterior–posterior axis by regulating *Hox* gene expression patterns during development (Jacobs and van Lohuizen 2002; Orland 2003). On the other hand, these two complexes play reciprocal roles in definitive hematopoiesis: negative regulation by the Eed-containing complex and positive regulation by Bmi-1-containing complex (Lessard et al. 1999). Several gene-targeting studies have revealed hematopoietic abnormalities suggesting involvement of PcG group proteins in hematopoiesis (Table 1). In Bmi-1 knock-out mice, progressive pancytopenia after birth, leading to death by 20 weeks of age, has been described (van der Lugt et al. 1994). *Mph1/Rae28*$^{-/-}$ fetal liver contains 20-fold fewer long-term lymphohematopoietic re-

Table 1. Hematopoietic defects in PcG gene-deficient mice

Drosophila	Mouse	Hematopoietic defects in KO mice
Polycomb (Pc)	M33	Hypoplastic thymus and spleen lymphopenia
Polyhomeotic (Ph)	Rae28/Mph1	Decreased HSC number in fetal liver Hypoplastic thymus and spleen
Posterior sex combs (Psc)	Bmi-1	Progressive pancytopenia, hypo-plastic thymus and spleen lymphopenia
	Mel-18	Hypoplastic thymus and spleen lymphopenia
Enhancer of Zeste [E(z)]	Enx1/Ezh2	Reduced rearrangement of the $V_H J558$
Extra sex combs (Esc)	Eed	Myelo- and lympho-Proliferative disease in hypomorphic mutants

populating HSCs than wild type (Ohta et al. 2002). All of these findings have uncovered novel aspects of stem cell regulation exerted by epigenetic modifications mediated by PcG proteins. However, the defects in HSCs in those mice has not yet been characterized in detail at the clonal level in vitro and in vivo.

In this study, both loss-of-function and gain-of-function analysis revealed a central role for Bmi-1, but not the other components, in the maintenance of HSC self-renewal (Iwama et al. 2004). Our findings indicate that the expression level of Bmi-1 is the critical determinant for the self-renewal capacity of HSCs.

6.2   Results

6.2.1   Expression of PcG Genes in Various Hematopoietic Cells

Expression analysis of *PcG* genes in human hematopoietic cells has demonstrated that *Bmi-1* is preferentially expressed in primitive cells, while other *PcG* genes, including *M33*, *Mel-18*, and *Mph1/Rae-28*, are not detectable in primitive cells but up-regulated along with differentiation (Lessard et al. 1998). Our detailed RT-PCR analysis of mouse hematopoietic cells, however, revealed that all *PcG* genes encoding components of the Bmi-1-containing complex, such as *Bmi-1*, *Mph1/Rae-28*, *M33*, and *Mel-18*, are highly expressed in CD34$^-$KSL HSCs that comprise only 0.004% of bone marrow mononuclear cells (Osawa et al. 1996), and all are down-regulated during differentiation in the bone marrow (BM) (Fig. 1a). In contrast, *Eed*, whose product composes another PcG complex, was ubiquitously expressed. These expression profiles support the idea of positive regulation of HSC self-renewal by the Bmi-1-containing complex (Park et al. 2003; Lessard et al. 2003). To evaluate the role of uncharacterized PcG components, Mel-18 and M33, in the maintenance of HSCs, we performed competitive repopulation assay using ten times more fetal liver cells from *Bmi-1$^{-/-}$*, *Mel-18$^{-/-}$*, or *M33$^{-/-}$* mice than competitor cells. As reported, *Bmi-1$^{-/-}$* fetal liver cells did not contribute at all to long-term reconstitution (Fig. 1b). Mel-18 is highly related to Bmi-1 in domain structure, particularly in their N-terminal Ring finger and helix-turn-helix domains. Unexpectedly, *Mel-18$^{-/-}$* fetal liver cells showed a very mild deficiency in repopulating capacity when compared to *Bmi-1$^{-/-}$* fetal liver cells (Fig. 1b). Moreover, *M33$^{-/-}$* fetal liver cells exhibited normal repopulating capacity in both primary (Fig. 1b) and secondary recipients (chimerism after 3 months; wild-type 91.6 ± 2.0 vs. *M33$^{-/-}$* 92.2 ± 3.4, *n*=4).

6.2.2   Defective Self-Renewal and Accelerated Differentiation of *Bmi-1$^{-/-}$* HSCs

A progressive postnatal decrease in the number of Thy1.1^{low}c-Kit$^+$Sca-1$^+$Lineage marker$^-$ HSCs has been observed in *Bmi-1$^{-/-}$* mice (Park et al. 2003). To evaluate the proliferative and differentiation capacity

a

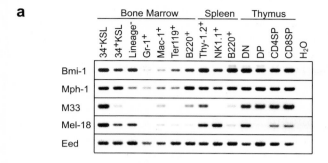

b

		E14.5 FL donor cells		Competitor Ly5.1 cells	PB chimerism 4 w	PB chimerism 12 w	
Bmi-1	+/+	2×10^6	(n=10)	2×10^5	90.2 ± 1.2	95.0 ± 1.2	
	+/–	2×10^6	(n=9)	2×10^5	86.1 ± 3.0	93.0 ± 2.4	*
	–/–	2×10^6	(n=10)	2×10^5	0.7 ± 0.5	0.4 ± 0.4	
Mel-18	+/+	2×10^6	(n=19)	2×10^5	86.1 ± 4.3	88.9 ± 2.9	*
	–/–	2×10^6	(n=7)	2×10^5	62.1 ± 5.0	64.9 ± 9.0	
M33	+/+	2×10^6	(n=3)	2×10^5	94.0 ± 2.3	98.9 ± 0.7	
	–/–	2×10^6	(n=2)	2×10^5	90.3	99.2	

*$p < 0.0001$

Fig. 1a,b. Role of components of the Bmi-1-containing complex in HSC. **a** mRNA expression of mouse *PcG* genes in hematopoietic cells. Cells analyzed are bone marrow CD34$^-$c-Kit$^+$Sca-1$^+$Lineage marker$^-$ stem cells (CD34$^-$KSL), CD34$^+$KSL progenitors, Lineage marker$^-$ cells (Lin$^-$), Gr-1$^+$ neutrophils, Mac-1$^+$ monocytes/macrophages, TER119$^+$ erythroblasts, B220$^+$B cells, spleen Thy-1.2$^+$ T cells, NK1.1$^+$ NK cells, B220$^+$ B cells, and thymic CD4$^-$CD8$^-$ T cells (DN), CD4$^+$CD8$^+$ T cells (DP), CD4$^+$CD8$^-$ T cells (CD4SP), and CD4$^-$CD8$^+$ (CD8SP). **b** Competitive lymphohematopoietic repopulating capacity of *PcG* gene-deficient HSCs. The indicated number of E14 fetal liver cells from *Bmi-1*$^{-/-}$, *Mel-18*$^{-/-}$, or *M33*$^{-/-}$ mice (B6-Ly5.2) and B6-Ly5.1 competitor cells were mixed and injected into lethally irradiated B6-Ly5.1 recipient mice. Percent chimerism of donor cells 4 and 12 weeks after transplantation is presented as mean $\pm$ SD

Fig. 2a–d. Defective self-renewal and accelerated differentiation of $Bmi-1^{-/-}$ HSCs. **a** Growth of $Bmi-1^{-/-}$ CD34⁻KSL HSCs in vitro. Freshly isolated CD34⁻KSL cells were cultured in the presence of SCF, IL-3, TPO, and EPO for 14 days. The results are shown as mean ± SD of triplicate cultures. **b** Single cell growth assay. Ninety-six individual CD34⁻KSL HSCs were sorted clonally into 96-well micro-titer plates in the presence of SCF, IL-3, TPO, and EPO. The numbers of high and low proliferative potential-colony-forming cells (HPP-CFC and LPP-CFC) were retrospectively evaluated by counting colonies at day 14 (HPP-CFC and LPP-CFC: colony diameter >1 mm and <1 mm, respectively). The results are shown as mean ± SD of triplicate cultures. **c** Frequency of each colony type. Colonies derived from HPP-CFC were recovered and morphologically examined for the composition of colony-forming cells. **d** Paired daughter assay. When a single CD34⁻KSL HSC underwent cell division and gave rise to two daughter cells, daughter cells were separated by micromanipulation and were further cultured to permit full differentiation along the myeloid lineage. The colonies were recovered for morphological examination

of $Bmi-1^{-/-}$ HSCs in BM, we purified the CD34⁻KSL HSC fraction, which is highly enriched for long-term repopulating HSCs (Osawa et al. 1996). $Bmi-1^{-/-}$ CD34⁻KSL cell showed comparable proliferation with wild-type and $Bmi-1^{+/--}$ cells for the first week of culture, but thereafter, they proliferated poorly (Fig. 2a). Single cell growth assays demonstrated that $Bmi-1^{-/-}$ CD34⁻KSL cells are able to form detectable colonies at a frequency comparable to $Bmi-1^{+/+}$ and $Bmi-1^{+/--}$ CD34⁻KSL cells, but contained threefold fewer high proliferative potential-colony-forming cells (HPP-CFCs). Reduction of HPP-CFCs that gave rise to colonies larger than 2 mm in diameter was even more prominent (sevenfold) (Fig. 2b). All HPP colonies larger than 1 mm in diameter were evaluated morphologically. Surprisingly, most of the HPP colonies generated from $Bmi-1^{-/-}$ CD34⁻KSL cells consisted of only neutrophils and macrophages. $Bmi-1^{-/-}$ CD34⁻KSL cells presented a ninefold reduction in their frequency of colony-forming unit-neutrophil/macrophage/Erythroblast/Megakaryocyte (CFU-nmEM), which retains multilineage differentiation capacity, compared with $Bmi-1^{+/+}$ CD34⁻KSL cells (Fig. 2c). Failure of $Bmi-1^{-/-}$ CD34⁻KSL cells to inherit multilineage differentiation potential through successive

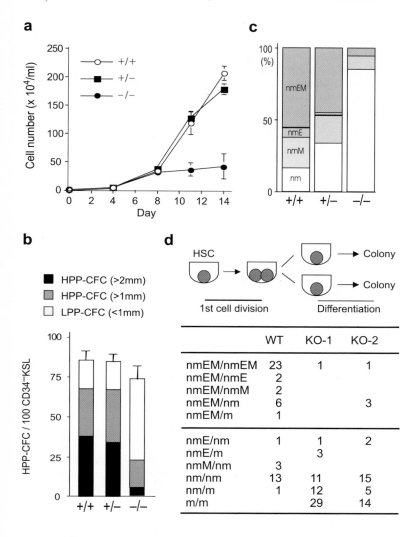

cell division was obvious in a paired daughter assay (Fig. 2d). In most daughter cell pairs generated from wild-type CD34⁻KSL cells, at least one of the two daughter cells inherit nmEM differentiation potential,

Fig. 3a–c. Rescue of defective $Bmi\text{-}1^{-/-}$ HSC function by re-expression of $Bmi\text{-}1$. **a** Wild-type and $Bmi\text{-}1^{-/-}$ CD34$^-$KSL cells were transduced with GFP control or $Bmi\text{-}1$ and plated in methylcellulose medium to allow colony formation 36 h after the initiation of transduction. GFP$^+$ colonies larger than 1 mm in diameter, which were derived from HPP-CFCs, were counted at day 14, and **b** recovered for morphological analysis to evaluate frequency of each colony type. The results are shown as mean $\pm$ SD of triplicate cultures. **c** Indicated numbers of $Bmi\text{-}1^{-/-}$ CD34$^-$KSL cells were transduced with $Bmi\text{-}1$. After 3.5 days from the initiation of transduction, cells were injected into lethally irradiated Ly5.1 recipient mice along with Ly5.1 competitor cells. Repopulation by rescued $Bmi\text{-}1^{-/-}$ CD34$^-$KSL cells was evaluated by monitoring donor cell chimerism in peripheral blood at the indicated time points after transplantation

whereas $Bmi\text{-}1^{-/-}$ CD34$^-$KSL cells showed accelerated loss of multilineage differentiation potential, leading to the limited differentiation and inefficient expansion of their progeny.

These profound defects of $Bmi\text{-}1^{-/-}$ HSC function evoke the possibility that absence of Bmi-1 in HSCs causes additional epigenetic abnormalities that are irreversible, and CD34$^-$KSL cells no longer retain stem cell properties. Retroviral transduction of $Bmi\text{-}1^{-/-}$ CD34$^-$KSL cells with $Bmi\text{-}1$, however, completely rescued their defects in proliferation and multilineage differentiation potential in vitro (Fig. 3a, b) and long-term repopulating capacity *in vivo* (Fig. 3d). These findings suggest that execution of stem cell activity is absolutely dependent on Bmi-1.

6.2.3 Augmentation of HSC Activity by Forced *Bmi-1* Expression

An essential role of Bmi-1 in the maintenance of HSC self-renewal capacity prompted us to determine augmentation of HSC activity by PcG genes. CD34$^-$KSL HSCs were transduced with $Bmi\text{-}1$, $Mph1/Rae28$, or $M33$, and then further incubated for 13 days (14-day ex vivo culture in total). Transduction efficiencies were over 80% in all experiments (data not shown). In the presence of SCF and TPO, which support expansion of HSCs and progenitors rather than their differentiation, forced expression of $Bmi\text{-}1$ as well as $Mph1/Rae28$ gave no apparent growth advantage in culture compared with the GFP control (Fig. 4a). Notably,

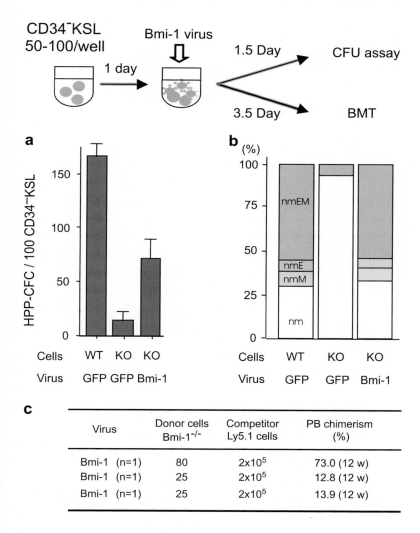

however, *Bmi-1*-transduced but not *Mph1/Rae28*-transduced cells contained numerous HPP-CFCs (Fig. 4b). Morphological evaluation of the colonies revealed significant expansion of CFU-nmEM by *Bmi-1*. Given

Fig. 4a–c. Ex vivo expansion of CFU-nmEM by forced expression of *Bmi-1* in HSCs. **a** CD34⁻KSL cells transduced with indicated *PcG* gene retroviruses were cultured in the presence of SCF and TPO and their growth was monitored. Morphology of cultured cells at day 14 was observed under an inverted microscope (*inset*). **b** At day 14 of culture, colony assays were performed to evaluate the content of HPP-CFC in culture. GFP⁺ colonies derived from HPP-CFCs were examined on their colony types by morphological analysis. **c** Net expansion of CFU-nmEM during the 14-day culture period. The results are shown as mean±SD of triplicate cultures

that 60% of freshly isolated CD34⁻KSL cells can be defined as CFU-nmEM, as shown in Fig. 2d, there was a net expansion of CFU-nmEM of 56- to 80-fold over 14 days in the *Bmi-1* cultures (Fig. 4c). Unexpectedly, expression of *M*33 induced an adverse effect on proliferation and caused accelerated differentiation into macrophages that attached the bottom of culture dishes (Fig. 4a). The effect of *Bmi-1* is comparable to that of *HoxB4*, a well-known HSC activator (Antonchuk et al. 2002) (Fig. 4c).

To determine the mechanism that leads to the drastic expansion of CFU-nmEM, which retains a full range of differentiation potential, we employed a paired daughter cell assay to see if overexpression of *Bmi-1* promotes symmetric HSC division in vitro. After 24-h prestimulation, CD34⁻KSL cells were transduced with a *Bmi-1* retrovirus for another 24 h. After transduction, single-cell cultures were initiated by micromanipulation. When a single cell underwent cell division, the daughter cells were separated again and were allowed to form colonies. To evaluate the commitment process of HSCs while excluding committed progenitors from this study, we selected daughter cells retaining nmEM differentiation potential by retrospective inference. Expression of *Bmi-1* was assessed by GFP expression. As expected, forced expression of *Bmi-1* significantly promoted symmetrical cell division of daughter cells (Fig. 5), indicating that Bmi-1 contributes to CFU-nmEM expansion by promoting self-renewal of HSCs

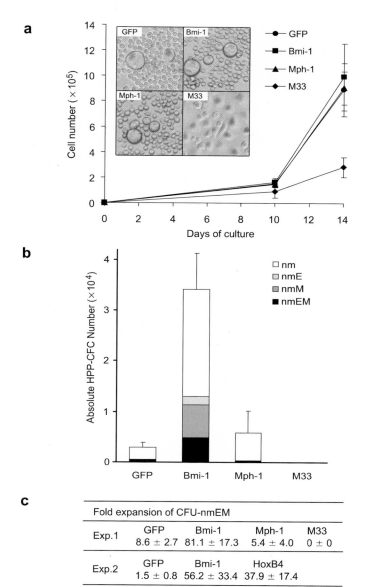

a

b

c

Fold expansion of CFU-nmEM			
Exp.1	GFP	Bmi-1	Mph-1	M33
	8.6 ± 2.7	81.1 ± 17.3	5.4 ± 4.0	0 ± 0
Exp.2	GFP	Bmi-1	HoxB4	
	1.5 ± 0.8	56.2 ± 33.4	37.9 ± 17.4	

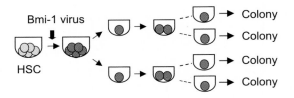

Colony pair	GFP virus	Bmi-1 virus
nmEM/nmEM	26 (53%)	40 (74%)*
nmEM/nmE	1	1
nmEM/nmM	3	6
nmEM/nm	19	7
Total	49	54

Fig. 5. Forced expression of *Bmi-1* promotes symmetrical cell division of HSCs. CD34⁻ KSL HSCs were transduced with either *GFP* or *Bmi-1* retroviruses. After 24 h of transduction, cells were separated clonally by micromanipulation. When a single cell underwent cell division, daughter cells were separated again by micromanipulation and were further cultured to permit full differentiation along the myeloid lineage. The colonies were recovered for morphological examination. Only the pairs whose parental cells should have retained neutrophil (*n*), macrophage (*m*), erythroblast (*E*), and Megakaryocyte (*M*) differentiation potential were selected. The probability of symmetrical cell division of daughter cells transduced with *Bmi-1* was significantly higher than the control ($p < 0.044$)

6.3   Discussion

Loss-of-function analyses of the *PcG* genes *Bmi-1* and *Mph1/Rae-28* have established that they are essential for the maintenance of adult BM HSCs, but not for the development of definitive HSCs (Ohta et al. 2002; Park et al. 2003; Lessard and Sauvageau 2003). Compared with *Mph1/Rae-28⁻/⁻* mice, however, hematopoietic defects are more severe in *Bmi-1⁻/⁻* mice and are attributed to impaired HSC self-renewal (Park et al. 2003; Lessard and Sauvageau 2003). In this study, although both *Mel-18* and *M33* genes appeared to be highly expressed in HSCs (Fig. 1a), *Mel-18⁻/⁻* and *M33⁻/⁻* HSCs showed mild or no defects and retained long-term repopulating capacity (Fig. 1b). Accordingly, overexpression of *PcG* genes in HSCs demonstrated that only *Bmi-1* enhances HSC function, while *M*33 completely abolishes HSC function (Fig. 4). All these findings clearly address a central role for Bmi-1 in the maintenance of HSCs and suggest that the level of Bmi-1 protein is a critical determinant for the activity of the PcG complex in HSCs. Bmi-1 may behave as a core component of the PcG complex in recruiting molecules essential for gene silencing, or provide a docking site for DNA-binding proteins that target PcG complex to polycomb response element (PRE). On the other hand, the finding that M33 is dispensable in the maintenance of definitive HSCs is surprising. Both *Bmi-1* and *M*33 are involved in the maintenance of homeotic gene expression pattern through development, and strong dosage interactions between the two genes have been observed in this process (Bel et al. 1998). Our finding, however, presents a possibility that M33 does not contribute to the Bmi-1 PcG complex in HSCs. M33 could be recruited to histone H3 Lysine 27 methylated by the Eed-containing complex and thereby mediate targeting of the Bmi-1-containing complex to PcG targets (Fischle et al. 2003). Thus, M33 is a key molecule for coordinated regulation of *Hox* genes by Eed- and Bmi-1-containing complexes. In contrast, the dispensable role of M33 in HSCs correlated well to the reciprocal roles of the two complexes in definitive hematopoiesis (Lessard et al., 1999) and indicates that Bmi-1-containing complex has a silencing pathway of its own. The negative effect of overloaded M33 on HSCs could be due to squelching of PcG components by M33.

HSCs are maintained and expanded through self-renewal. HSC self-renewal secures its high repopulation capacity and multilineage differentiation potential through cell division. If HSCs fail to self-renew, they differentiate to lower orders of progenitors with limited proliferative and differentiation potential. Paired daughter cell assays that monitor the behavior of HSCs in vitro (Suda et al. 1984; Takano et al. 2004) demonstrated that Bmi-1 is essential for CD34⁻KSL cells to inherit multilineage differentiation potential through successive cell divisions (Fig. 2d). Notably, overexpression of *Bmi-1* in CD34⁻KSL cells promoted their symmetrical cell division, indicating a higher probability of inheritance of stemness mediated by *Bmi-1* (Fig. 5). This is the first evidence of successful genetic manipulation of HSC self-renewal in vitro.

These clonal observations together with functional rescue of $Bmi-1^{-/-}$ HSC both in vitro and in vivo strongly support an essential role of Bmi-1 in HSC self-renewal.

The central role for Bmi-1 in HSC self-renewal was also demonstrated by overexpression experiments of *PcG* genes in HSCs. The *Bmi-1*-mediated growth advantage was largely restricted to the primitive hematopoietic cells. During ex vivo culture, total cell numbers were almost comparable to the control while a net 56- to 80-fold CFU-nmEM expansion was obtained in the *Bmi-1* cultures (Fig. 4). In agreement with these data, symmetrical cell division of HSCs was promoted in the *Bmi-1* cultures (Fig. 5), suggesting enhanced probability of HSC self-renewal and progenitor expansion mediated by *Bmi-1* overexpression. The comparable effect of *Bmi-1* to that of *HoxB4*, a well-known HSC activator (Antonchuk et al. 2002), is noteworthy. Recent findings indicated that genetic manipulation of *HoxB4* can support generation of long-term repopulating HSCs from ES cells (Kyba et al. 2002), and ex vivo expansion of HSCs can be obtained by direct targeting of HoxB4 protein into HSCs (Amsellem et al. 2003; Krosl et al. 2003). Similar to HoxB4, Bmi-1 could be a novel target for therapeutic manipulation of HSCs. Although PcG proteins regulate expression of homeotic genes including *HoxB4* during development (Takihara et al. 1997), deregulation of *Hox* genes in definitive hematopoietic cells has not yet been identified in mice deficient for PcG genes (Ohta et al. 2002; Park et al. 2003; Lessard et al. 2003). However, the enhancement of HSC activity by two genes is highly similar in many aspects. It will be intriguing to ask how these two molecules work as HSC activators.

The mechanism whereby Bmi-1 maintains HSCs remains to be defined. Although derepression of Bmi-1 target genes $p16$ and $p19$ has been attributed to defective HSC self-renewal, the cell cycle status of $CD34^-KSL$ HSCs was not grossly altered in $Bmi-1^{-/-}$ mice (data not shown). Therefore, a detailed analysis of $Bmi-1^{-/-}p16^{-/-}p19^{-/-}$ HSCs will be necessary to define their roles in HSCs. One attractive hypothesis is that derepression of $p16$ and $p19$ genes causes early senescence of primitive hematopoietic cells as reported in $Bmi-1^{-/-}$ mouse embryonic fibroblasts (Jacobs et al. 1999). On the other hand, our preliminary data indicates that there exist additional Bmi-1 targets other than $p16$ and $p19$ that are essential for HSC self-renewal. Further analysis of Bmi-1 target genes would clarify a novel aspect of the regulatory mechanism of HSC self-renewal.

References

Amsellem S, Pflumio F, Bardinet D, Izac B, Charneau P, Romeo P-H, Dubart-Kupperschmitt A, Fichelson S (2003) Ex vivo expansion of human hematopoietic stem cells by direct delivery of the HoxB4 homeoprotein. Nat Med 9:1423–1427

Antonchuk J, Sauvageau G, Humphries RK (2002) HOXB4-induced expansion of adult hematopoietic stem cells ex vivo. Cell 109:39–45

Bel S, Core N, Djabali M, Kieboom K, van der Lugt N, Alkema MJ, van Lohuizen M (1998) Genetic interactions and dosage effects of polycomb group genes in mice. Development 125:3543–3551

Fischle W, Wang Y, Jacobs SA, Kim Y, Allis CD, Khorasanizadeh S (2003) Molecular basis for the discrimination of repressive methyl-lysine marks in histone H3 by polycomb HP1 chromodomains. Genes Dev 17:1870–1881

Iwama A, Oguro H, Negishi M, Kato Y, Morira V, Tsukui M, Ema H, Kamijo T, Katoh-Fukui Y, Koseki H, Lohuizen van M, and Nakauchi H. (2004) Enhanced self-renewal of hematopoietic stem cells mediated by the polycomb gene product, Bmi-1. Immunity 21:843–851

Jacobs JJL, Kieboom K, Marino S, DePinho RA, van Lohuizen M (1999) The oncogene polycomb-group gene Bmi1 regulates proliferation and senescence through the ink4a locus. Nature 397:164–168

Jacobs JJL, van Lohuizen M (2002) Polycomb repression: from cellular memory to cellular proliferation and cancer. Biochim Biophys Acta 1602:151–161

Krosl J, Austin P, Beslu N, Kroon E, Humphries RK, Sauvageau G (2003) In vitro expansion of hematopoietic stem cells by recombinant TAT-HOXB4 protein. Nat Med 9:1428–1432

Kyba M, Perlingeiro RCR Daley GQ (2002) HoxB4 confers definitive lymphoid-myeloid engraftment potential on embryonic stem cell and yolk sac hematopoietic progenitors. Cell 109:29–37

Lessard L, Baban S, Sauvageau G (1998) Stage-specific expression of polycomb group genes in human bone marrow cells. Blood 91:1216–1224

Lessard J, Schumacher A, Thorsteinsdottir U, van Lohuizen M, Magnuson T, Sauvageau Guy (1999) Functional antagonism of the polycomb-group genes Eed Bmi1 in hematopoietic cell proliferation. Genes Dev 13:2691–2703

Lessard J, Sauvageau G (2003) Bmi-1 determines proliferative capacity of normal and leukemic stem cells. Nature 423:255–260

Ohta H, Sawada A, Kim JY, Tokimasa S, Nishiguchi S, Humphries RK, Hara J, Takihara Y (2002) Polycomb group gene rae28 is required for sustaining activity of hematopoietic stem cells. J Exp Med 195:759–770

Orland V (2003) Polycomb, epigenomes, and control of cell identity. Cell 112:599–606

Osawa M, Hanada K-I, Hamada H, Nakauchi H (1996) Long-term lympho-hematopoietic reconstitution by a single CD34-low/negative hematopoietic stem cells. Science 273:242–245

Park I-K, Qian D, Kiel M, Becker MW, Pihalja M, Weissman IL, Morrison SJ, Clarke MF (2003) Bmi-1 is required for maintenance of adult self-renewing haematopoietic stem cells. Nature 423:302–305

Suda T, Suda J, Ogawa M (1984) Disparate differentiation in mouse hematopoietic colonies derived paired progenitors. Proc Natl Acad Sci U S A 81:2520–2524

Takano H, Ema H, Sudo K, Nakauchi H (2004) Asymmetric division and lineage commitment at the level of hematopoietic stem cells: inference from differentiation in daughter cell and granddaughter cell pairs. J Exp Med 199:295–302

Takihara Y, Tomotsune D, Shirai M, Katoh-Fukui Y, Nishi K, Motaleb MA, Nomura M, Tsuchiya R, Fujita Y, Shibata Y, Higashinakagawa T, Shimoda K (1997) Targeted disruption of the mouse homologue of the Drosophila polyhomeotic gene leads to altered anteroposterior patterning and neural crest defects. Development 124:3673–3682

Van der Lugt NM, Domen J, Linders K, van Room M, Robanus-Maandag E, te Riele H, van der Valk M, Deschamps J, Sofroniew M, van Lohuizen M (1994) Posterior transformation, neurological abnormalities, and severe hematopoietic defects in mice with a targeted deletion of the Bmi-1 proto-oncogene. Genes Dev 8:757–769

Wang H, Wang L, Erdjument-Bromage H, Vidal M, Tempst P, Jones RS, Zhang Y (2004) Role of histone H2A ubiquitination in polycomb silencing. Nature 431:873–877

7 Directed Differentiation of Neural and Sensory Tissues from Embryonic Stem Cells In Vitro

Y. Sasai

Abstract. We have recently identified a stromal cell-derived inducing activity (SDIA), which induces differentiation of neural cells from mouse embryonic stem (ES) cells. Particularly, midbrain TH^+ dopaminergic neurons are generated efficiently in this system. These dopaminergic neurons are transplantable and survive well in the 6-OHDA-treated mouse striatum. SDIA induces co-cultured ES cells to differentiate into rostral central nervous system (CNS) tissues containing both ventral and dorsal cells. While early exposure of SDIA-treated ES cells to BMP4 suppresses neural differentiation and promotes epidermogenesis, late BMP4 exposure after the 4th day of co-culture causes differentiation of neural crest cells and dorsal-most CNS cells, with autonomic system and sensory lineages induced preferentially by high and low BMP4 concentrations, respectively. In contrast, Sonic Hedgehog (Shh) suppresses differentiation of neural crest lineages and promotes that of ventral CNS tissues such as motor neurons and $HNF3\beta^+$ floor plate cells with axonal guidance activities. Thus,

SDIA-treated ES cells generate naïve precursors that have the competence of differentiating into the "full" dorsal–ventral range of neuroectodermal derivatives in response to patterning signals. I also discuss the role of SDIA and the mode of rostral-caudal specification of neuralized ES cells. In addition, I would like to discuss them in the light of control of in vitro neural production for the use in regenerative medicine for parkinsonism and retinal degeneration.

7.1   Introduction

In vertebrate embryogenesis, the primordia of the nervous systems arise from uncommitted ectoderm during gastrulation. Spemann and Mangold (1924) demonstrated that the dorsal lip of the amphibian blastopore, which gives rise mainly to axial mesoderm, emanates inductive factors that direct neural differentiation in ectoderm. The nature of these neural inducers has represented a "holy grail" in the field of developmental biology for several decades. Over the last several years, molecular studies in *Xenopus* have identified neural inducer molecules and revealed their mode of action (Sasai and De Robertis 1996). Neural inducers such as noggin and Chordin (Lamb et al. 1993; Sasai et al. 1995) induce neural differentiation in isolated *Xenopus* ectoderm (animal caps) and promote dorsalization of mesoderm when acting on mesodermal precursors. These neural inducers do not have their own receptors on target cells, instead they act by binding to and inactivating BMP4, which suppresses neural differentiation and ventralizes mesoderm (Wilson and Hemmati-Brivanlou 1995; Sasai and De Robertis 1996). These results suggest that both neural induction and mesoderm dorsalization are controlled by a common morphogenetic signaling, that is, a BMP activity gradient (Sasai 2000).

By contrast, relatively little is known about regulatory factors in mammalian neural induction. One main reason for this is that good experimental systems for in vitro neural differentiation are still lacking in mice, that is, something comparable to the animal cap assay commonly used in *Xenopus* studies. Mammalian ES cells can differentiate into all embryonic cell types when injected into blastocyst-stage embryos. This pluripotency of ES cells can be partially recapitulated in vitro by floating culture of ES cell aggregates, or embryoid bodies (EBs). After a few weeks of culture without LIF, EBs frequently contain ectodermal, mesodermal, and endodermal derivatives. However, there have not been any good methods that can induce "selective" differentiation to a particular cell type such as neurons. Therefore, we attempted to develop such an experimental system by using ES cell culture.

7.2 Neural Differentiation in ES Cells Using the SDIA Method

We first asked whether attenuation of BMP signaling is sufficient to induce neural differentiation of mouse ES cells by administering BMP antagonist molecules (Kawasaki et al. 2000). Neither transfection of pCMV-Chordin plasmid into ES cells nor addition of neutralizing BMPR-Fc receptobody to culture medium induced significant neural differentiation of ES cells. On the other hand, administration of BMP4 protein efficiently suppressed in vitro neural differentiation of mouse ES cells (e.g., EB+RA method) or of isolated mouse epiblast even at a low concentration (0.5 nM). These results indicate that blockade of BMP4 signaling is required but not sufficient for neural differentiation of undifferentiated mouse cells. This suggests that mouse ES cells, unlike *Xenopus* cells, require some unknown signals for initiating neural differentiation of ES cells, in addition to attenuation of BMP signals.

By using a coculture system, we screened various primary culture cells and cell lines for activities promoting neural differentiation of ES cells under serum-free conditions (Kawasaki et al. 2000). Some stromal (or mesenchymal) cells promoted neural differentiation when used as feeders. PA6 cells (stromal cells derived from skull bone marrow)(Kodama et al. 1986) induced remarkably efficient neural differentiation when cocultured with ES cells, resulting in 92% of colonies becoming NCAM-positive by day 12.

In these colonies, the majority of cells were stained with either the neuronal marker TuJ (class III β-tubulin) or the neural precursor marker nestin. Very few colonies contained GFAP-positive cells (2%). The TuJ-positive neurons also expressed other neuronal markers such as MAP2 and neurofilament, and the presynaptic marker synaptophysin was detected on the induced neurons. In contrast to the high NCAM-positive percentage, very few colonies expressed mesodermal markers such as

PDGFR alpha, Flk1 and MF20 (all <2% colonies). Thus, PA6 can promote neural differentiation of cocultured ES cells without inducing mesodermal markers. We named the neural-inducing activity on the stromal cells "stromal cell-derived inducing activity" or "SDIA"(Kawasaki et al. 2000).

At present, the molecular nature of SDIA remains to be identified. One important question that arises is whether PA6-derived factors act by antagonizing BMP4 in a similar manner as Chordin and noggin. We actively blocked BMP signaling by administering neutralizing BMPR-Fc receptobody, and found that the presence of BMPR-Fc did not affect the extent of neural differentiation in the ES cells under any conditions tested. Thus, BMP antagonism is unlikely to explain the neuralizing activity of SDIA. One possible explanation is that some additional factors such as SDIA are required for neural differentiation before mouse cells make the neural/epidermal binary decision in a BMP-dependent manner. Consistent with this idea, the SDIA-treated cells acquire their highest sensitivity to BMP4 subsequent to the onset of nestin expression (day 3). This indicates that SDIA has already exerted some effects (nestin induction) before the cells react to BMP signals. The following scenario might be applicable to the mechanism of neuralization occurring in SDIA-treated ES cells. First, ES cells cultured on PA6 move in an ectodermal direction under the influence of SDIA. SDIA-treated ES cells then adopt a default neural status unless they receive a sufficient level of BMP4 signals. However, as the molecular nature of SDIA remains to be elucidated, we must await further study to judge this proposition and to understand the relevant roles of SDIA in the embryo.

7.3   Induction of Dopaminergic Neurons by SDIA Culture

Immunohistochemical analyses of the characteristics of SDIA-induced neurons revealed that TH neurons occupied 30% of TuJ-positive neurons (~16% of total cells; Kawasaki et al. 2000). This value is again significantly higher than percentages of GABAergic, cholinergic, and serotonergic neurons in TuJ-positive neurons (18%, 9%, and 2%, respectively) at the cell level. A time course study showed that TH-positive neurons appeared between days 6–8 of the induction period. The cells remain negative for DBH (dopamine-β-hydroxylase; marker for norepinephrine and epinephrine neurons) even after 13 days of induction. The mesencephalic dopaminergic neuron markers Nurr1 and Ptx3 were induced in SDIA-treated ES cells. HPLC analyses showed that ES cell-derived neurons released a significant amount of dopamine into the medium (7.7 pmol/10^6 cells) in response to a depolarizing stimulus. These data show that functional neurons producing dopamine were generated with this method.

The SDIA method is technically simple and the induction is efficient and speedy. Studies with various differentiation markers have demonstrated that time course features of neural differentiation by SDIA in vitro mimic well those observed in the developing embryo. The SDIA method does not involve EB formation or RA treatment, and each differentiating colony grows from a single ES cell in two dimensions under serum-free conditions. Because of these features, the SDIA method has advantages over the EB/RA methods (Bain et al. 1995) and the five-step induction/selection method (Lee et al. 2000) when used for detailed analyses of differentiation, such as effects of exogenous growth factors.

7.4   Application to Cell Transplantation

We tested whether SDIA-treated ES cells could be integrated into the mouse striatum after implantation (Kawasaki et al. 2000; Morizane et al. 2002). ES cell colonies were cultured on PA6 cells for 12 days and detached en bloc ($\sim$50 μm) from the feeders by a mild protease treatment without EDTA. The isolated ES cell colonies were then implanted into the mouse striatum, which had been treated with 6-hydroxydopamine (6-OHDA). Ipsilateral implantation of SDIA-induced neurons significantly restored TH-positive areas in and around the DiI-positive graft. Two weeks after implantation of 4×10^5 SDIA-treated ES cells, 3.9×10^4 grafted cells were found in the brain and 74% of them were TuJ-positive neurons. The estimated survival rate of TH-positive neurons was approximately 22% after these procedures. No teratoma formation was observed in the grafted tissue by histology

In collaboration with Dr. Jun Takahashi, Kyoto University Hospital, we performed preclinical translational research toward clinical application of SDIA-induced dopaminergic neurons to parkinsonian therapeutics by using primate (Macaca) ES cells (Suemori et al. 2001; Kawasaki et al. 2002). In this study, we generated neurospheres composed of neural progenitors from monkey ES cells, which are capable of producing large numbers of DA neurons. We are currently analyzing the effect of transplantation of DA neurons generated from monkey ES cells into MPTP-treated monkeys, a primate model for Parkinson's disease. Behavioral studies and functional imaging will be examined in detail to see whether the transplanted cells function as DA neurons and attenuated MPTP-induced neurological symptoms.

7.5 Differentiation of Other Neural Tissues

To understand the range of competence of ES cell-derived neural precursors induced by SDIA, we examined in vitro differentiation of mouse and primate ES cells into the dorsal-most (neural crest) and the ventral-most (floor plate) cells of the neural axis (Mizuseki et al. 2003). SDIA induces cocultured ES cells to differentiate into rostral CNS tissues containing both ventral and dorsal cells. While early exposure of SDIA-treated ES cells to BMP4 suppresses neural differentiation and promotes epidermogenesis, late BMP4 exposure after the 4th day of coculture causes differentiation of neural crest cells and dorsal-most CNS cells. Autonomic neuron precursors and sensory neuron lineages are induced preferentially by high and low BMP4 concentrations, respectively. In contrast, Shh suppresses differentiation of neural crest lineages and promotes that of ventral CNS tissues such as motor neurons. Notably, high concentrations of Shh efficiently promote differentiation of $HNF3\beta^+$ floor plate cells ($\sim$30% total cells) with axonal guidance activities. Thus, SDIA-treated ES cells generate naïve precursors that have the competence of differentiating into the "full" dorsal–ventral range of neuroectodermal derivatives in response to patterning signals.

7.6   Induction of Eye Tissues from ES Cells

Unexpectedly, we observed in vitro production of retinal pigment epithelium from ES cells. In addition to TH$^+$ cells, 3%–8% of the colonies contain large patches of Pax6$^+$ pigmented epithelium (Kawasaki et al. 2002). These primate ES cell-derived pigment epithelial cells exhibit the morphologic (electron microscopy), biochemical (marker proteins), and functional characteristics (phagocytosis and tight junction formation) expected of the retinal pigment epithelium. In collaboration with Dr. Masayo Takahashi (Kyoto University Hospital), we succeeded in culturing and expanding pigment epithelia on the matrix-coated dish. The sheet of cultured pigment epithelium cells is uniform and ready for transplantation. When transplanted subretinally into the rat model of photoreceptor degeneration (RCS rat; Li and Turner 1988) caused by dysfunction of the retinal pigment epithelium, these pigment epithelial cells enhance the survival of the photoreceptor cells (Haruta et al. 2004). Furthermore, the rotation drum test has shown that the implantation of these ES cell-derived pigment epithelial cells rescue visual functions of the RCS rat. The present study raises the possibility that embryonic stem cells constitute a source of cells for retinal transplantation in patients with dysfunction of the retinal pigment epithelium. Dysfunction of the retinal pigment epithelium is associated with ocular diseases such as age-related macular degeneration and retinitis pigmentosa, which are among the leading causes of blindness. The SDIA method provides an unlimited source of primate cells for the study of these retinal diseases.

When cynomologus monkey ES cells are induced to differentiate by SDIA, we found that lens differentiation occurs occasionally after long culture (Ooto et al. 2003). Following a 4-week induction period, lentoids are produced by a subpopulation of ES colonies cultured on PA6 cells. Western blotting and immunohistochemistry revealed that these lentoids expressed alpha-crystallin and Pax6. The number of lentoids is increased with increasing FGF-2 concentration and plated colony density (up to 20%–30% colonies). It is intriguing that both neural and placodal derivatives of the eye (retinal cells and lens) are induced in SDIA-treated ES cells. We are currently optimizing the conditions for lens differentiation.

7.7 Conclusion and Future Perspectives

In summary, the SDIA system serves as a versatile in vitro system for neural differentiation of ES cells to be used for basic and applied stem cell research. For further development of ES cell-based therapeutics, it is essential to establish techniques for preparation of clinically "safe" cells. This includes:

1. Purification system of the cells of interest (e.g., dopaminerfic neuron progenitors)
2. Removal of hazardous cells such as tumor-forming cells
3. GMP-class culture system
4. Control of rejection
5. Replacement of PA6 cells with human-derived stromal cells

References

Bain G, Kitchens D, Yao M, Huettner JE, Gottlieb DI (1995). Embryonic stem cells express neuronal properties in vitro. Dev Biol 168:342–357

Haruta M, Sasai Y, Kawasaki H, Amemiya K, Ooto S, Kitada M, Suemori H, Nakatsuji N, Ide C, Honda Y, Takahashi M (2004) In vitro and in vivo characterization of pigment epithelial cells differentiated from primate embryonic stem cells. Invest Ophthalmol Vis Sci 45:1020–1025

Kawasaki H, Mizuseki K, Sasai Y (2001) Selective neural induction from ES cells by stromal cell-derived inducing activity and its possible application to parkinsonian therapy. In: Turksen K (ed) Methods in molecular biology. Humana Press, Totowa NJ, 217–228

Kawasaki H, Suemori H, Mizuseki K, Watanabe K, Urano F, Ichinose H, Haruta M, Takahashi M, Yoshikawa K, Nishikawa SI, Nakatsuji N, Sasai Y (2002) Generation of TH^+ dopaminergic neurons, $Pax6^+$ pigment epithelia from primate ES cells by SDIA. PNAS 99:1580–1585

Kawasaki H, Mizuseki K, Nishikawa S, Kaneko S, Kuwana Y, Nakanishi S, Nishikawa S-I, Sasai Y (2000) Induction of midbrain dopaminergic neurons from ES cells by stromal cell-derived inducing activity. Neuron 28:31–40

Kodama H, Hagiwara H, Sudo H, Amagai Y, Yokota T, Arai N, Kitamura Y (1986) MC3T3-G2/PA6 preadipocytes support in vitro proliferation of hemopoietic stem cells through a mechanism different from that of interleukin 3. J Cell Physiol 129:20–26

Lamb TM, Knecht AK, Smith WC, Stachel SE, Economides AN, Stahl N, Yancopolous GD, Harland RM (1993) Neural induction by the secreted polypeptide noggin. Science 262:713–718

Lee S-H, Lumelsky N, Studer L, Auerbach JM, McKay RD (2000). Efficient generation of midbrain and hindbrain neurons from mouse embryonic stem cell. Nature Biotech 18:675–679

Li LX, Turner JE (1988) Inherited retinal dystrophy in the RCS rat: prevention of photoreceptor degeneration by pigment epithelial cell transplantation. Exp Eye Res 47:911–917

Mizuseki K, Sakamoto T, Watanabe K, Muguruma K, Ikeya M, Nishiyama A, Arakawa A, Suemori H, Nakatsuji N, Kawasaki H, Murakami F, Sasai Y (2003) Generation of neural crest-derived PNS neurons, floor plate cells from mouse, primate ES Cells. Proc Natl Acad Sci U S A 100:5828–5833

Morizane A, Takahashi J, Takagi Y, Sasai Y, Hashimoto N (2002) Optimal conditions for in vivo induction of dopaminergic neurons from embryonic stem cells through stromal cell-derived inducing activity. J Neurosci Res 69934–69939

Ooto S, Haruta M, Honda Y, Kawasaki H, Sasai Y, Takahashi M (2003) Induction of the differentiation of lentoids from primate embryonic stem cells. Invest Ophthalmol Vis Sci 44:2689–2693

Sasai Y, De Robertis EM (1997) Ectodermal patterning in vertebrate embryos. Dev Biol 182:5–20

Sasai Y, Lu B, Steinbeisser H, De Robertis EM (1995) Regulation of neural induction by the chd, BMP-4 antagonistic patterning signals in *Xenopus*. Nature 376:333–336

Sasai Y (2000) Regulation of neural determination by evolutionarily conserved signals: anti-BMP factors and what next? Curr Opin Neurobiol 11:22–26

Spemann H, Mangold H (1924) Über induktion von Embryonalagen durch Implantation Artfremder Organisatoren. Roux' Arch Entw Mech 100:599–638

Suemori H, Tada T, Torii R et al. (2001) Establishment of embryonic stem cell lines from cynomolgus monkey blastocysts produced by IVF or ICSI. Dev Dyn. 222:273–279

Wilson PA, Hemmati-Brivanlou A (1995) Induction of epidermis and inhibition of neural fate by BMP-4. Nature 376:331–333

8 Stem Cell Biology for Vascular Regeneration

T. Asahara

Abstract. The isolation of endothelial progenitor cells (EPCs) derived from bone marrow (BM) was one epoch-making event for the recognition of neovessel formation in adults occurring as physiological and pathological responses. The finding that EPCs home to sites of neovascularization and differentiate into endothelial cells (ECs) in situ is consistent with vasculogenesis, a critical paradigm that has been well described for embryonic neovascularization, but proposed recently in adults in which a reservoir of stem or progenitor cells contribute to vascular organogenesis. EPCs have also been considered as therapeutic agents to supply the potent origin of neovascularization under pathological conditions. This chapter highlights an update of EPC biology as well as its potential use for therapeutic regeneration.

8.1   Introduction

Tissue regeneration by somatic stem/progenitor cell has been recognized as a maintenance or recovery system of many organs in the adult. The isolation and investigation of these somatic stem/progenitor cells has described how these cells contribute to postnatal organogenesis. On the basis of the regenerative potency, these stem/progenitor cells are expected to be a key strategy of therapeutic applications for damaged organs.

Recently endothelial progenitor cells (EPCs) have been isolated from adult peripheral blood (PB). EPCs are considered to share common stem/progenitor cells with hematopoietic stem cells and have been shown to derive from bone marrow (BM) and to incorporate into foci of physiological or pathological neovascularization. The finding that EPCs home to sites of neovascularization and differentiate into endothelial cells (ECs) in situ is consistent with vasculogenesis, a critical paradigm that has been well described for embryonic neovascularization, but recently proposed in adults in which a reservoir of stem/progenitor cells contribute to postnatal vascular organogenesis. The discovery of EPCs has therefore drastically changed our understanding of adult blood vessel formation. This chapter provides the update of EPC biology and highlights the potential utility of EPCs for therapeutic vascular regeneration.

8.2   Postnatal Neovascularization

By the discovery of EPCs in PB (Asahara et al. 1997; Shi et al. 1998), our understanding of postnatal neovascularization has been expanded from angiogenesis to angio/vasculogenesis. As previously described (Folkman and Shing 1992), postnatal neovascularization was originally recognized to be constituted by the mechanism of angiogenesis, which is neovessel formation, operated by in situ proliferation and migration of preexisting endothelial cells. However, the isolation of EPCs resulted in the addition of the new mechanism, vasculogenesis, which is de novo vessel formation by in situ incorporation, differentiation, migration, and/or proliferation of BM-derived EPCs (Asahara et al. 1999a) (Fig. 1). More recently, tissue-specific stem/progenitor cells with the potency of differentiation into myocytes or ECs was isolated in skeletal muscle tissue of murine hindlimb, although the origin remains to be cleared (Tamaki et al. 2002)]. This finding suggests that the origin of EPCs may not be limited to BM, e.g., tissue-specific stem/progenitor cells possibly provide in situ EPCs as other sources of EPCs than BM.

In the events of minor-scale neovessel formation, i.e., slight wounds or burns, in situ preexisting ECs causing postnatal angiogenesis may replicate and replace the existing cell population enough, as ECs exhibit the ability for self-repair that preserves their proliferative activity. Neovascularization through differentiated ECs, however, is limited in terms

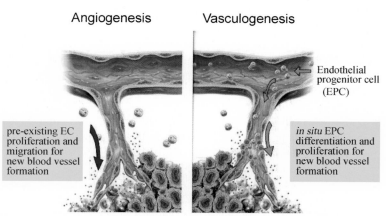

Fig. 1. Postnatal neovascularization in the physiological or pathological events is consistent with neovessel formation contributed by angiogenesis and vasculogenesis at the various rates between their two mechanisms. Angiogenesis and vasculogenesis are due to the activations of in situ ECs and BM-derived or in situ EPCs, respectively

of cellular life span (the Hayflick limit) and their inability to incorporate into remote target sites. In the case of large-scale tissue repair, such as in patients who experienced acute vascular insult secondary to burns, coronary artery bypass grafting (CABG), or acute myocardial infarction (Gill et al. 2001; Shintani et al. 2001b), or in physiological cyclic organogenesis of endometrium Asahara et al. 1999a), BM-derived or in situ EPC kinetics are activated under the influence of appropriate cytokines, hormones, and/or growth factors through the autocrine, paracrine, and/or endocrine systems. Thus the contemporary view of tissue regeneration is that neighboring differentiated ECs are relied upon for vascular regeneration during a minor insult, whereas tissue-specific or BM-derived stem/progenitor cells bearing EPCs/ECs are important when an emergent vascular regenerative process is required (Fig. 1).

8.3   Profiles of EPCs in Adults

8.3.1   The Isolation of EPCs in Adults

In the embryo, evidence suggests that hematopoietic stem cells (HSCs) and EPCs (Risau et al. 1988; Pardanaud et al. 1987) are derived from a common precursor (hemangioblast) (Flamme and Risau 1992; Weiss and Orkin 1996). During embryonic development, multiple blood islands initially fuse to form a yolk sac capillary network (Risau and Flamme 1995), which provides the foundation for an arteriovenous vascular system that eventually forms following the onset of blood circulation (Risau et al. 1988). The integral relationship between the cells that circulate in the vascular system (the blood cells) and those principally responsible for the vessels themselves (ECs) is suggested by their spatial orientation within the blood islands; those cells destined to generate hematopoietic cells are situated in the center of the blood island (HSCs) versus EPCs or angioblasts, which are located at the periphery of the blood islands. In addition to this arrangement, HSCs and EPCs share certain common antigens, including CD34, KDR, Tie-2, CD117, and Sca-1 (Choi et al. 1998).

The existence of HSCs in the PB and BM, and the demonstration of sustained hematopoietic reconstitution with HSC transplantation led to an idea that a closely related cell type, namely EPCs, may also exist in adult tissues. Recently, EPCs were successfully isolated from circulating mononuclear cells (MNCs) using KDR, CD34, and CD133 antigens shared by both embryonic EPCs and HSCs (Asahara et al. 1997; Yin et al. 1997; Peichev et al. 2000). In vitro, these cells differentiate into endothelial lineage cells, and in animal models of ischemia, heterologous, homologous, and autologous EPCs have been shown to incorporate into the foci of neovasculature, contributing to neovascularization. Recently, similar studies with EPCs isolated from human cord blood have demonstrated their analogous differentiation into ECs in vitro and in vivo (Nieda et al. 1997; Murohara et al. 2000); Kang et al. 2001; Crisa et al. 1999).

These findings have raised important questions regarding fundamental concepts of blood vessel growth and development in adult subjects. Does the differentiation of EPCs in situ (vasculogenesis) play an important role in adult neovascularization, and would impairments in this process lead to clinical diseases? There is now a strong body of evidence suggesting that vasculogenesis does in fact make a significant contribution to postnatal neovascularization. Recent studies with animal BM transplantation (BMT) models in which BM (donor)-derived EPCs could be distinguished have shown that the contribution of EPCs to neovessel formation may range from 5% to 25% in response to granulation tissue formation (Crosby et al. 2000) or growth factor-induced neovascularization (Murayama et al. 2002). Also, in tumor neovascularization, the range is approximately 35%–45% higher than the former events (Reyes et al. 2002). The degree of EPC contribution to postnatal neovascularization is predicted to depend on each neovascularizing event or disease.

8.3.2 Diverse Identifications of Human EPCs and Their Precursors

Since the initial report of EPCs (Asahar et al. 1997; Shi et al. 1998), a number of groups have set out to better define this cell population. Because EPCs and HSCs share many surface markers, and no simple definition of EPCs exists, various methods of EPC isolation have been reported (Asahar et al. 1997; Shi et al. 1998; Piechev et al. 2000; Nieda et al. 1997; Murohara et al. 2000; Kang et al. 2001; Boyer et al. 2000; Lin et al. 2000; Kalka et al. 2000a; Gunsilius et al. 2000; Gehling et al. 2000; Fernandez Pujol et al. 2000; Schatteman et al. 2000; Harraz et al. 2001; Quirici et al. 2001). The term "EPC" may therefore encompass a group of cells that exist in a variety of stages ranging from hemangioblasts to fully differentiated ECs. Although the true differentiation lineage of EPCs and their putative precursors remain to be determined, there is overwhelming evidence in vivo that a population of EPCs exists in human.

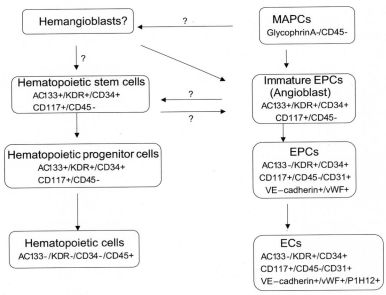

Fig. 2. Origin and differentiation of EPCs. EPCs are thought to differentiate not only from putative hemangioblasts, common precursor cells with HSCs, as previously described, but also from MAPCs. Representative antigenicities to stem/progenitor cells are shown (+; positive, −; negative)

Lin et al. cultivated peripheral MNCs from patients receiving gender-mismatched BMT and studied their growth in vitro. In this study, they identified a population of BM (donor)-derived ECs with high proliferative potential (late outgrowth); these BM cells likely represent EPCs (Lin et al. 2000). Gunsilius, et al. investigated a chronic myelogenous leukemia model and disclosed that BM-derived EPCs contribute to postnatal neovascularization in human (Gunsilius et al. 2000). Interestingly, in the report, BM-derived EPCs could be detected even in the wall of quiescent vessels without neovascularization events. The finding suggests that BM-derived EPCs may be related even to the turnover of ECs consisting of quiescent vessels.

Reyes et al. have recently isolated multipotent adult progenitor cells (MAPCs) from BM MNCs, differentiated them into EPCs and proposed MAPCs as an origin of EPCs (Reyes et al. 2002). These studies therefore provide evidence to support the presence of BM-derived EPCs that take part in neovascularization. Also, as described above, the existence of namely in situ EPCs as derived from tissue-specific stem/progenitor cells in murine skeletal muscle remains to be investigated even in the other tissues (Tamaki et al. 2002) (Fig. 2).

8.4   EPC Kinetics in Adults

8.4.1   EPC Kinetics Effected by Endogenous Agents

The incorporation of BM-derived EPCs into foci of physiological and pathological neovascularization has been demonstrated through various animal experiments. One well-established model that allows the detection of BM-derived EPCs includes transplanting wild-type mice with BM cells harvested from transgenic mice in which LacZ expression is regulated by an EC lineage-specific promoter, flk-1 or Tie-2 (flk-1/lacZ/BMT, Tie-2/lacZ/BMT). Using such mice, flk-1- or Tie-2-expressing endothelial lineage cells derived from BM (EPCs) have been shown to localize to vessels during tumor growth, wound healing, skeletal and cardiac ischemia, cornea neovascularization, and endometrial remodeling following hormone-induced ovulation (Asahara et al. 1999a).

Tissue trauma causes mobilization of hematopoietic cells as well as pluripotent stem or progenitor cells from the hematopoietic system (Grzelak et al. 1998), as indicated by past research. Consistent with the notion that EPCs and HSCs share a common ancestry, recent data from our laboratory have shown that mobilization of BM-derived EPCs constitutes a natural response to tissue ischemia. The aforementioned murine BMT model also provided direct evidence of enhanced BM-derived EPC incorporation into foci of corneal neovascularization following the development of hindlimb ischemia (Takahashi et al. 1999). This finding indicates that circulating EPCs are mobilized endogenously in response to tissue ischemia and can incorporate into neovascular foci to promote tissue repair. These results in animals were recently confirmed by human studies illustrating EPC mobilization in patients following burns (Gill et al. 2001), CABG, or acute myocardial infarction (Shintani et al. 2001b).

As previous studies demonstrated, the role of endogenous mobilization of BM-derived EPCs, we considered exogenous mobilization of EPCs as an effective means of augmenting the resident population of EPCs/ECs. Such a strategy is appealing for its potential to overcome the endothelial dysfunction or depletion that may be associated with older, diabetic, or hypercholesterolemic patients. Granulocyte macrophage colony-stimulating factor (GM-CSF) is well known to stimulate hematopoietic progenitor cells and myeloid lineage cells, but has recently been shown to exert a potent stimulatory effect on EPC kinetics. The delivery of this cytokine induced EPC mobilization and enhanced neovascularization of severely ischemic tissues and de novo corneal vascularization (Takahashi et al. 1999).

The exact mechanism by which EPCs are mobilized to the peripheral circulation remains unknown, but may mimic aspects of embryonic development. Vascular endothelial growth factor (VEGF), critical for angio/vasculogenesis in the embryo (Shalaby et al. 1995; Carmeliet et al. 1996; Ferrara et al. 1996), has recently been shown to be an important stimulus of adult EPC kinetics. Our studies performed first in mice (Asahara et al. 1999b) and subsequently in patients undergoing VEGF gene transfer for critical limb or myocardial ischemia [38] established a previously unappreciated mechanism by which VEGF contributes to neovascularization in part by mobilizing BM-derived EPCs. Similar modulation of EPC kinetics has been observed in response to other hematopoietic stimulators, such as granulocyte-colony stimulating factor (G-CSF) and stroma-derived factor-1 (SDF-1) (Moore et al. 2001).

8.4.2 EPC Kinetics Effected by Exogenous Agents

EPC mobilization has recently been implicated not only by natural hematopoietic or angiogenic stimulants but also by pharmacological agents. For instance, 3-hydroxy-3-methylglutaryl coenzyme A (HMG-CoA) reductase inhibitors (statins) are known to rapidly activate Akt signaling in ECs, thereby stimulating EC bioactivity in vitro and enhancing angiogenesis in vivo (Kureishi et al. 2000). Recent works by Dimmeler et al. and our laboratory have demonstrated a novel function of statins by mobilizing BM-derived EPCs through the stimulation of the Akt signaling pathway (Llevadot et al. 2001; Dimmeler et al. 2001; Vasa et al. 2001a; Urbich et al. 2002). Therefore this newly appreciated role of statins, along with their already well-established safety and efficacy on hypercholesterolemia, suggests that they can offer benefit in treating various forms of vascular diseases. On the other hand, some antiangiogenic agents, i.e., angiostatin or soluble flk-1, are shown to inhibit BM-derived EPC kinetics, leading to tumor regression (Davidoff et al. 2001), as BM-derived EPC kinetics is a critical factor for tumor growth, in terms of tumor neovascularization (Lyden et al. 2001).

8.4.3 Clinical Profile of EPC Kinetics

There is a body of strong evidence to suggest that impaired neovascularization results in part from diminished cytokine production. However, endogenous expression of cytokines is not the only factor leading to impaired neovascularization. Diabetic or hypercholesterolemic animals – like clinical patients – exhibit evidence of dysfunction in mature endothelial cells. While the cellular dysfunction does not necessarily preclude a favorable response to cytokine replacement therapy, the extent of recovery in limb perfusion in these animals fails to reach that of control animals; this suggests another limitation imposed by a diminished responsiveness of EPCs/ECs (Rivard et al. 1999b; Van Belle et al. 1997; Couffinhal et al. 1999).

The aging characterized by impaired neovascularization (Rivard et al. 1999a, 2000) might be associated with dysfunctional EPCs and defective vasculogenesis. Indeed, preliminary results from our laboratory indicate that transplantation of BM (including EPCs) from old mice into young mice led to minimal neovascularization in a corneal micropocket assay, relative to transplantation of young BM. We also demonstrated that EPCs from older patients with clinical ischemia had significantly less therapeutic effect in rescuing ischemic hindlimb of mice compared with those from younger ischemic patients (Murayama et al. 2001). These studies provide evidence to support an age-dependent impairment in vasculogenesis (as well as angiogenesis) that is heavily influenced by EPC phenotype. Moreover, analysis of clinical data from older patients at our institution disclosed a significant reduction in the number of EPCs at baseline, as well as that in response to VEGF165 gene transfer (Kalka et al. 2000b). Thus impaired EPC mobilization and/or activity in response to VEGF may contribute to the age-dependent defect in postnatal neovascularization. Recently Vasa et al. have further investigated EPC kinetics and their relationship to clinical disorders, showing that the number and migratory activity of circulating EPCs inversely correlates with risk factors for coronary artery disease, such as smoking, family history, and hypertension (Vasa et al. 2001b). On the basis of these findings, monitoring of BM-derived EPC kinetics in the patients with vascular diseases is expected to be valuable in the evaluation of lesion activity and/or therapeutic efficacy.

8.5 Therapeutic Vasculogenesis

8.5.1 The Potential of EPC Transplantation

The regenerative potential of stem/progenitor cells is currently under intense investigation. In vitro, stem/progenitor cells possess the capability of self-renewal and differentiation into organ-specific cell types. When placed in vivo, these cells are then provided with the proper milieu that allows them to reconstitute organ systems. The novel strategy of EPC transplantation (cell therapy) may therefore supplement the classic paradigm of angiogenesis developed by Folkman and colleagues. Our studies indicated that cell therapy with culture-expanded EPCs can successfully promote neovascularization of ischemic tissues, even when administered as sole therapy, i.e., in the absence of angiogenic growth factors. Such a supply-side version of therapeutic neovascularization in which the substrate (EPCs/ECs) rather than ligand (growth factor) comprises the therapeutic agent, was first demonstrated by intravenously transplanting human EPCs to immunodeficient mice with hindlimb ischemia (Kalka et al. 2000a). These findings provided novel evidence that exogenously administered EPCs rescue impaired neovascularization in an animal model of critical limb ischemia. Not only did the heterologous cell transplantation improve neovascularization and blood flow recovery, but it also led to important biological outcomes, notably, the reduction of limb necrosis and auto-amputation by 50% in comparison with controls. A similar strategy applied to a model of myocardial ischemia in the nude rat demonstrated that transplanted human EPCs localize to areas of myocardial neovascularization, differentiate into mature ECs, and enhance neovascularization. These findings were associated with preserved left ventricular (LV) function and diminished myocardial fibrosis (Kawamoto et al. 2001). Murohara et al. reported similar findings in which human cord blood-derived EPCs also augmented neovascularization in hindlimb ischemic model of nude rats, followed by in situ transplantation (Murohara et al. 2000).

Other researchers have more recently explored the therapeutic potential of freshly isolated human $CD34^+$ MNCs (EPC-enriched fraction). Shatteman et al. conducted local injection of freshly isolated human CD34+ MNCs into diabetic nude mice with hindlimb ischemia, and showed an increase in the restoration of limb flow (Shatteman et al. 2000). Similarly Kocher et al. attempted intravenous infusion of freshly isolated human CD34+ MNCs into nude rats with myocardial ischemia and found preservation of LV function associated with inhibition of cardiomyocyte apoptosis (Kocher et al. 2001). Thus two approaches of EPC preparation (i.e., both cultured and freshly isolated human EPCs) may provide therapeutic benefit in vascular diseases, but as described below, will likely require further optimization techniques to acquire the ideal quality and quantity of EPCs for EPC therapy.

8.5.2   Future Strategy of EPC Cell Therapy

Ex vivo expansion of EPCs cultured from PB-MNCs of healthy human volunteers typically yields 5.0×10^6 cells per 100 ml of blood on
day 7. Our animal studies (Kalka et al. 2000a) suggest that heterologous transplantation requires systemic injection of $0.5–2.0 \times 10^4$ human
EPCs per gram body weight of the recipient animal to achieve satis-
factory reperfusion of an ischemic hindlimb. Rough extrapolation of
these data to human suggests that a blood volume of as much as 12 l
may be necessary to obtain adequate numbers of EPCs to treat critical limb ischemia in patients. Therefore, the fundamental scarcity of
EPCs in the circulation, combined with their possible functional impairment associated with a variety of phenotypes in clinical patients, such
as aging, diabetes, hypercholesterolemia, and homocyst(e)inemia (vide
infra), constitute major limitations of primary EPC transplantation. Considering autologous EPC therapy, certain technical improvements that
may help to overcome the primary scarcity of a viable, and the functional
EPC population should include:

1. Local delivery of EPCs
2. Adjunctive strategies (e.g., growth factor supplements) to promote
   BM-derived EPC mobilization (Takahashi et al. 1999; Asahara et al.
   1999b)
3. Enrichment procedures, i.e., leukapheresis or BM aspiration
4. Enhancement of EPC function by gene transduction (gene modified
   EPC therapy, vide infra)
5. Culture expansion of EPCs from self-renewable primitive stem cells
   in BM or other tissues

Alternatively, unless the quality and quantity of autologous EPCs to sat-
isfy the effectiveness of EPC therapy can be acquired by the technical
improvements described above, allogenic EPCs derived from umbilical cord blood or culture-expanded from human embryonic stem cells
(Murohara et al. 2000; Levenberg 2002) may be available as the sources
supplying EPCs.

8.5.3 Gene-Modified EPC Therapy

A strategy that may alleviate potential EPC dysfunction in ischemic disorders is considered reasonable, given the findings that EPC function and mobilization may be impaired in certain disease states. Genetic modification of EPCs to overexpress angiogenic growth factors, to enhance signaling activity of the angiogenic response, and to rejuvenate the bioactivity and/or extend the life span of EPCs can constitute such potential strategies.

We have recently shown for the first time that gene-modified EPCs rescue impaired neovascularization in an animal model of limb ischemia (Iwaguro et al. 2002). Transplantation of heterologous EPCs transduced with adenovirus encoding human VEGF165 not only improved neovascularization and blood flow recovery, but also had meaningful biological consequences, i. e., limb necrosis and autoamputation were reduced by 63.7% in comparison with controls. Notably, the dose of EPCs needed to achieve limb salvage in these in vivo experiments was 30 times less than that required in the previous experiments involving unmodified EPCs (Kalka et al. 2000a). Thus, combining EPC cell therapy with gene (i.e., VEGF) therapy may be one option to address the limited number and function of EPCs that can be isolated from peripheral blood in patients.

8.5.4 BM-MNC Transplantation

Nonselected total BM cells or BM-MNCs including an immature EPC population have also been investigated for their potential to induce neovascularization. Several experiments have reported that autologous BM administration into the rabbit (Shintani et al. 2001a) or rat (Hamano et al. 2001) hindlimb ischemic model and the porcine myocardial ischemic model (Kamihata et al. 2001; Fuchs et al. 2001) could augment neovascularization in ischemic tissue mainly through the production of angiogenic growth factors and less through the differentiation of a portion of the cells into EPCs/ECs in situ. Although there are no long-term safety and efficacy data for local delivery of such cell populations mostly composed of inflammatory leukocytes, these strategies have already been applied to clinical patients in some institutions and preliminary results are expected soon.

8.6   Other Potential of EPCs

EPCs have recently been applied to the field of tissue engineering as a means of improving biocompatibility of vascular grafts. Artificial grafts first seeded with autologous CD34$^+$ cells from canine BM and then implanted into the aortae were found to have increased surface endothelialization and vascularization compared with controls (Bhattacharya et al. 2000). Similarly, when cultured autologous ovine EPCs were seeded onto carotid interposition grafts, the EPC-seeded grafts achieved physiological motility and remained patent for 130 days vs 15 days in nonseeded grafts (Kaushal et al. 2001). Alternatively, as previously reported, the cell sheets of cultured cardiomyocytes may be effective for the improvement of cardiac function in the damaged hearts, i.e., ischemic heart disease or cardiomyopathy (Shimizu et al. 2002a, 2002b). The cell sheets consisting of cardiomyocytes with EPCs expectedly inducing neovessels may be attractive, as blood supply is essential to maintain the homeostasis of implanted cardiomyocytes in such cell sheets.

8.7   Conclusion

As the concept of BM-derived EPCs in adults and postnatal vasculogenesis are further established, clinical applications of EPCs to regenerative medicine are likely to follow. To acquire the more optimized quality and quantity of EPCs, several issues remain to be addressed, such as the development of more efficient methods of EPC purification and expansion, the methods of administration and senescence in EPCs. Alternatively, in case autologous BM-derived EPCs cannot be used in the patients with impaired BM function, appreciable EPCs isolated from umbilical cord blood or differentiated from tissue-specific stem/progenitor or embryonic stem cells need to be optimized for EPC therapy. However, the unlimited potential of EPCs along with the emerging concepts of autologous cell therapy with gene modification suggest that they may soon reach clinical fruition.

References

Asahara T, Murohara T, Sullivan A et al (1997) Isolation of putative progenitor endothelial cells for angiogenesis. Science 275:964–967

Asahara T, Masuda H, Takahashi T et al (1999a) Bone marrow origin of endothelial progenitor cells responsible for postnatal vasculogenesis in physiological and pathological neovascularization. Circ Res 85:221–228

Asahara T, Takahashi T, Masuda H et al (1999b) VEGF contributes to postnatal neovascularization by mobilizing bone marrow-derived endothelial progenitor cells. EMBO J 18:3964–3972

Bhattacharya V, McSweeney PA, Shi Q et al (2000) Enhanced endothelialization and microvessel formation in polyester grafts seeded with CD34(+) bone marrow cells. Blood 95:581–585

Boyer M, Townsend LE, Vogel LM et al (2000) Isolation of endothelial cells and their progenitor cells from human peripheral blood. J Vasc Surg 31:181–189

Carmeliet P, Ferreira V, Breier G et al (1996) Abnormal blood vessel development and lethality in embryos lacking a single VEGF allele. Nature 380:435–439

Choi K, Kennedy M, Kazarov A et al (1998) A common precursor for hematopoietic and endothelial cells. Development 125:725–732

Couffinhal T, Silver M, Kearney M et al (1999) Impaired collateral vessel development associated with reduced expression of vascular endothelial growth factor in ApoE–/– mice. Circulation 99:3188–3198

Crisa L, Cirulli V, Smith KA et al (1999) Human cord blood progenitors sustain thymic T-cell development and a novel form of angiogenesis. Blood 94:3928–3940

Crosby JR, Kaminski WE, Schatteman G et al (2000) Endothelial cells of hematopoietic origin make a significant contribution to adult blood vessel formation. Circ Res 87:728–730

Davidoff AM, Ng CY, Brown P et al (2001) Bone marrow-derived cells contribute to tumor neovasculature and, when modified to express an angiogenesis inhibitor, can restrict tumor growth in mice. Clin Cancer Res 7:2870–2879

Dimmeler S, Aicher A, Vasa M et al (2001) HMG-CoA reductase inhibitors (statins) increase endothelial progenitor cells via the PI 3-kinase/Akt pathway. J Clin Invest 108:391–397

Fernandez Pujol B, Lucibello FC et al (2000) Endothelial-like cells derived from human CD14 positive monocytes. Differentiation 65:287–300

Ferrara N, Carver Moore K, Chen H et al (1996) Heterozygous embryonic lethality induced by targeted inactivation of the VEGF gene. Nature 380:439–442

Flamme I, Risau W (1992) Induction of vasculogenesis and hematopoiesis in vitro. Development 116:435–439

Folkman J, Shing Y (1992) Angiogenesis. J Biol Chem 267:10931–10934

Gill M, Dias S, Hattori K et al (2001) Vascular trauma induces rapid but transient mobilization of VEGFR2(+)AC133(+) endothelial precursor cells. Circ Res 88:167–174

Fuchs S, Baffour R, Zhou YF et al (2001) Transendocardial delivery of autologous bone marrow enhances collateral perfusion and regional function in pigs with chronic experimental myocardial ischemia. J Am Coll Cardiol 37:1726–1732

Gehling UM, Ergun S, Schumacher U et al (2000) In vitro differentiation of endothelial cells from AC133-positive progenitor cells. Blood 95:3106–3112

Grzelak I, Olszewski WL, Zaleska M et al (1998) Surgical trauma evokes a rise in the frequency of hematopoietic progenitor cells and cytokine levels in blood circulation. Eur Surg Res 30:198–204

Gunsilius E, Duba HC, Petzer AL et al (2000) Evidence from a leukaemia model for maintenance of vascular endothelium by bone-marrow-derived endothelial cells. Lancet 355:1688–1691

Hamano K, Li TS, Kobayashi T et al (2001) The induction of angiogenesis by the implantation of autologous bone marrow cells: a novel and simple therapeutic method. Surgery 130:44–54

Harraz M, Jiao C, Hanlon HD et al (2001) CD34- blood-derived human endothelial cell progenitors. Stem Cells 19:304–312

Iwaguro H, Yamaguchi J, Kalka C et al (2002) Endothelial progenitor cell vascular endothelial growth factor gene transfer for vascular regeneration. Circulation 105:732–738

Kalka C, Masuda H, Takahashi T et al (2000a) Transplantation of ex vivo expanded endothelial progenitor cells for therapeutic neovascularization. Proc Natl Acad Sci U S A 97:3422–3427

Kalka C, Masuda H, Takahashi T et al (2000b) Vascular endothelial growth factor(165) gene transfer augments circulating endothelial progenitor cells in human subjects. Circ Res 86:1198–1202

Kamihata H, Matsubara H, Nishiue T et al (2001) Implantation of bone marrow mononuclear cells into ischemic myocardium enhances collateral perfusion and regional function via side supply of angioblasts, angiogenic ligands, and cytokines. Circulation 104:1046–1052

Kang HJ, Kim SC, Kim YJ et al (2001) Short-term phytohaemagglutinin-activated mononuclear cells induce endothelial progenitor cells from cord blood CD34+ cells. Br J Haematol 113:962–969

Kaushal S, Amiel GE, Guleserian KJ et al (2001) Functional small-diameter neovessels created using endothelial progenitor cells expanded ex vivo. Nat Med 7:1035–1040

Kawamoto A, Gwon HC, Iwaguro H et al (2001) Therapeutic potential of ex vivo expanded endothelial progenitor cells for myocardial ischemia. Circulation 103:634–637

Kocher AA, Schuster MD, Szabolcs MJ et al (2001) Neovascularization of ischemic myocardium by human bone-marrow-derived angioblasts prevents cardiomyocyte apoptosis, reduces remodeling and improves cardiac function. Nat Med 7:430–436

Kureishi Y, Luo Z, Shiojima I et al (2000) The HMG-CoA reductase inhibitor simvastatin activates the protein kinase Akt and promotes angiogenesis in normocholesterolemic animals. Nat Med 6:1004–1010

Levenberg S, Golub JS, Amit M et al (2002) Endothelial cells derived from human embryonic stem cells. Proc Natl Acad Sci U S A 99:4391–4396

Lin Y, Weisdorf DJ, Solovey A et al (2000) Origins of circulating endothelial cells and endothelial outgrowth from blood. J Clin Invest 105:71–77

Llevadot J, Murasawa S, Kureishi Y et al (2001) HMG-CoA reductase inhibitor mobilizes bone marrow-derived endothelial progenitor cells. J Clin Invest 108:399–405

Lyden D, Hattori K, Dias S et al (2001) Impaired recruitment of bone-marrow-derived endothelial and hematopoietic precursor cells blocks tumor angiogenesis and growth. Nat Med 7:1194–1201

Moore MA, Hattori K, Heissig B et al (2001) Mobilization of endothelial and hematopoietic stem and progenitor cells by adenovector-mediated elevation of serum levels of SDF-1, VEGF, and angiopoietin-1. Ann N Y Acad Sci 938:36–45; discussion 45–47

Murayama T, Kalka C, Silver M et al (2001) Aging impairs therapeutic contribution of human endothelial progenitor cells to postnatal neovascularization. Abstract Circulation 104:2–68

Murayama T, Tepper OM, Silver M et al (2002) Determination of bone marrow-derived endothelial progenitor cell significance in angiogenic growth factor-induced neovascularization in vivo. Exp Hematol 30:967–972

Murohara T, Ikeda H, Duan J et al (2000) Transplanted cord blood-derived endothelial precursor cells augment postnatal neovascularization. J Clin Invest 105:1527–1536

Nieda M, Nicol A, Denning Kendall P et al (1997) Endothelial cell precursors are normal components of human umbilical cord blood. Br J Haematol 98:775–777

Pardanaud L, Altmann C, Kitos P et al (1987) Vasculogenesis in the early quail blastodisc as studied with a monoclonal antibody recognizing endothelial cells. Development 100:339–349

Peichev M, Naiyer AJ, Pereira D et al (2000) Expression of VEGFR-2 and AC133 by circulating human CD34(+) cells identifies a population of functional endothelial precursors. Blood 95:952–958

Quirici N, Soligo D, Caneva L et al (2001) Differentiation and expansion of endothelial cells from human bone marrow CD133(+) cells. Br J Haematol 115:186–194

Reyes M, Dudek A, Jahagirdar B et al (2002) Origin of endothelial progenitors in human postnatal bone marrow. J Clin Invest 109:337–346

Risau W, Flamme I (1995) Vasculogenesis. Annu Rev Cell Dev Biol 11:73–91

Risau W, Sariola H, Zerwes HG et al (1988) Vasculogenesis and angiogenesis in embryonic-stem-cell-derived embryoid bodies. Development 102:471–478

Rivard A, Fabre JE, Silver M et al (1999a) Age-dependent impairment of angiogenesis. Circulation 99:111–120

Rivard A, Silver M, Chen D et al (1999b) Rescue of diabetes-related impairment of angiogenesis by intramuscular gene therapy with adeno-VEGF. Am J Pathol 154:355–363

Rivard A, Berthou Soulie L, Principe N et al (2000) Age-dependent defect in vascular endothelial growth factor expression is associated with reduced hypoxia-inducible factor 1 activity. J Biol Chem 275:29643–29647

Shalaby F, Rossant J, Yamaguchi TP et al (1995) Failure of blood-island formation and vasculogenesis in Flk-1-deficient mice. Nature 376:62–66

Shatteman GC, Hanlon HD, Jiao C et al (2000) Blood-derived angioblasts accelerate blood-flow restoration in diabetic mice. J Clin Invest 106:571–578

Shi Q, Rafii S, Wu MH et al (1998) Evidence for circulating bone marrow-derived endothelial cells. Blood 92:362–367

Shimizu T, Yamato M, Akutsu T et al (2002a) Electrically communicating three-dimensional cardiac tissue mimic fabricated by layered cultured cardiomyocyte sheets. J Biomed Mater Res 60:110–117

Shimizu T, Yamato M, Isoi Y et al (2002b) Fabrication of pulsatile cardiac tissue grafts using a novel 3-dimensional cell sheet manipulation technique and temperature-responsive cell culture surfaces. Circ Res 90:e40

Shintani S, Murohara T, Ikeda H et al (2001a) Augmentation of postnatal neo-vascularization with autologous bone marrow transplantation. Circulation 103:897–903

Shintani S, Murohara T, Ikeda H et al (2001b) Mobilization of endothelial progenitor cells in patients with acute myocardial infarction. Circulation 103:2776–2779

Tamaki T, Akatsuka A, Ando K et al (2002) Identification of myogenic-endothelial progenitor cells in the interstitial spaces of skeletal muscle. J Cell Biol 157:571–577

Urbich C, Dernbach E, Zeiher AM et al (2002) Double-edged role of statins in angiogenesis signaling. Circ Res 90:737–744

Van Belle E, Rivard A, Chen D et al (1997) Hypercholesterolemia attenuates angiogenesis but does not preclude augmentation by angiogenic cytokines. Circulation 96:2667–2674

Vasa M, Fichtlscherer S, Adler K et al (2001a) Increase in circulating endothelial progenitor cells by statin therapy in patients with stable coronary artery disease. Circulation 103:2885–2890

Vasa M, Fichtlscherer S, Aicher A et al (2001b) Number and migratory activity of circulating endothelial progenitor cells inversely correlate with risk factors for coronary artery disease. Circ Res 89:E1–E7

Weiss MJ, Orkin SH (1996) In vitro differentiation of murine embryonic stem cells. New approaches to old problems. J Clin Invest 97:591–595

Yin AH, Miraglia S, Zanjani ED et al (1997) AC133, a novel marker for human hematopoietic stem and progenitor cells. Blood 90:5002–5012

Ernst Schering Research Foundation Workshop

Editors: Günter Stock
Monika Lessl

Vol. 33 (2001): Stem Cells from Cord Blood, In Utero Stem Cell
Development, and Transplantation-Inclusive Gene Therapy
Editors: W. Holzgreve, M. Lessl

Vol. 34 (2001): Data Mining in Structural Biology
Editors: I. Schlichting, U. Egner

Vol. 35 (2002): Stem Cell Transplantation and Tissue Engineering
Editors: A. Haverich, H. Graf

Vol. 36 (2002): The Human Genome
Editors: A. Rosenthal, L. Vakalopoulou

Vol. 37 (2002): Pharmacokinetic Challenges in Drug Discovery
Editors: O. Pelkonen, A. Baumann, A. Reichel

Vol. 38 (2002): Bioinformatics and Genome Analysis
Editors: H.-W. Mewes, B. Weiss, H. Seidel

Vol. 39 (2002): Neuroinflammation – From Bench to Bedside
Editors: H. Kettenmann, G.A. Burton, U. Moenning

Vol. 40 (2002): Recent Advances in Glucocorticoid Receptor Action
Editors: A. Cato, H. Schaecke, K. Asadullah

Vol. 41 (2002): The Future of the Oocyte
Editors: J. Eppig, C. Hegele-Hartung

Vol. 42 (2003): Small Molecule-Protein Interaction
Editors: H. Waldmann, M. Koppitz

Vol. 43 (2003): Human Gene Therapy:
Present Opportunities and Future Trends
Editors: G.M. Rubanyi, S. Ylä-Herttuala

Vol. 44 (2004): Leucocyte Trafficking:
The Role of Fucosyltransferases and Selectins
Editors: A. Hamann, K. Asadullah, A. Schottelius

Vol. 45 (2004): Chemokine Roles in Immunoregulation and Disease
Editors: P.M. Murphy, R. Horuk

Vol. 46 (2004): New Molecular Mechanisms of Estrogen Action
and Their Impact on Future Perspectives in Estrogen Therapy
Editors: K.S. Korach, A. Hillisch, K.H. Fritzemeier

Vol. 47 (2004): Neuroinflammation in Stroke
Editors: U. Dirnagl, B. Elger

Vol. 48 (2004): From Morphological Imaging to Molecular Targeting
Editors: M. Schwaiger, L. Dinkelborg, H. Schweinfurth

Vol. 49 (2004): Molecular Imaging
Editors: A.A. Bogdanov, K. Licha